T0235823

Springer Biographies

The books published in the Springer Biographies tell of the life and work of scholars, innovators, and pioneers in all fields of learning and throughout the ages. Prominent scientists and philosophers will feature, but so too will lesser known personalities whose significant contributions deserve greater recognition and whose remarkable life stories will stir and motivate readers. Authored by historians and other academic writers, the volumes describe and analyse the main achievements of their subjects in manner accessible to nonspecialists, interweaving these with salient aspects of the protagonists' personal lives. Autobiographies and memoirs also fall into the scope of the series.

Friedel Weinert

Karl Popper

Professional Philosopher and Public
Intellectual

 Springer

Friedel Weinert
Faculty of Social Sciences
University of Bradford
Bradford, UK

ISSN 2365-0613   ISSN 2365-0621 (electronic)
Springer Biographies
ISBN 978-3-031-15426-3   ISBN 978-3-031-15424-9 (eBook)
https://doi.org/10.1007/978-3-031-15424-9

This Springer imprint is published by the registered company Springer Nature Switzerland AG
The registered company address is: Gewerbestrasse 11, 6330 Cham, Switzerland

I dedicate this book to Brigitte, Eileen and Adrien

Preface

This book is the result of an invitation from Angela Lahee, my editor at Springer, to write a contribution to their new series on Springer Biographies. I had some familiarity with Popper's work but I have since learnt how little I knew of the complexity and subtlety of Popper's ideas. I set about interviewing people who knew Popper personally or work in his area of expertise. I met and discussed aspects of Popper's work with Steven French, Greg Raddick and Jon Hodge of the University of Leeds. I had lunch with David Miller (Warwick), his former assistant, in Birmingham and coffee with John Worrall at the London School of Economics (*LSE*). Ian Jarvie (Toronto), Jeremy Shearmur (Dumfries) and Mauricio Suàrez (Madrid) answered some of my questions via email.

Several people read the whole or parts of the draft manuscript and gave me linguistic, stylistic and factual advice: John Ackroyd (Bradford University); Mauricio Suarez of Complutense University, Madrid. I am indebted to John Worrall (*LSE*) who provided me with great insight into Popper's life at the *LSE* and urged me to clarify several arguments and choose better examples. But my greatest debt goes to David Miller who subjected the manuscript to close scrutiny. In the process he corrected many factual and linguistic mistakes, made stylistic suggestions, gave me valuable references and challenged me to clarify many of the claims I make. Thanks to David's advice it became a more accurate account of Popper's life and work. He also urged me to read Popper's extensive correspondence with many of the leading intellectuals of his day. I therefore owe a great thank you to Dr. Thomas Hainscho, of the Popper Sammlung in Klagenfurt, for his advice and help. He responded quickly to my frequent requests for digitalized files of Popper's correspondence and answered my queries. The perusal of his numerous letters has given me valuable insight into Popper's life, personality and work.

I thank them all. My approach, namely to see Popper's work as attempted solutions to the problems he perceived, has led me to emphasize aspects of his work—like his views on realism, testability and time—which have not, I believe, been sufficiently stressed in other intellectual biographies. This approach produced an interpretation of Popper's ideas, with which some will disagree. If any mistakes remain, they are entirely my own responsibility. This book can be read selectively. Readers who are less interested in the details of, say, his views on quantum mechanics or evolutionary theory can skip these chapters without losing sight of the overall picture. Finally, I would like to thank Angela Lahee for her unwavering support.

Bradford, UK   Friedel Weinert

Contents

Chapter 1
Introduction

The present volume is an intellectual and social biography of Sir Karl Raimund Popper, one of the most important philosophers of the twentieth century. He was born in Vienna on July 28, 1902, and died in the Mayday Hospital in Croydon, near London, on September 17, 1994. The book introduces the reader to Popper's intellectual work and social life. Many of his ideas reached well beyond philosophy. The question is why do they matter today? Biographies usually follow a chronological order. As I wanted to discuss and evaluate Popper's ideas on a wide range of political, social and scientific issues, I have chosen a different approach: it combines chronology and analysis. I have applied Popper's *method of the problem situation*, also known as *situational analysis*, to his own work. That means, I present Popper's philosophical ideas as tentative solutions to perceived problems, against a background of previous solutions and techniques. I follow the evolution of his thematic ideas within each period of his productive life; but keep an eye on their development in his later career. The narrative is interwoven with details of his social life—his awards, honours, impacts, friendships and feuds—largely derived from his correspondence.

   As some of his commentators have observed, an underlying unity runs through his thought. (Watkins 1974; Lecourt 1994; O'Hear 1995: 2) It lies in the objectivity of his approach: his opposition to subjectivism; his emphasis on problems and tentative solutions; the critical attitude towards authority—be it in science, politics or society. Popper himself placed the unity of his philosophy on the importance of Critical Rationalism, that is the abandonment of all interest in the traditional (Cartesian) justification of knowledge claims. ('Replies' 1974: 1053) He professed to be one of the last 'laggards of the Enlightenment'. (See, for instance, *Offene Gesellschaft* ³1983: 20; *Realism/Aim* 1983/1985: 177; *Alles Leben* 1996: 158; *All Life* 1999: 83) Popper was not a religious man; he described himself as 'agnostic'. (*Self/Brain* 1977: viii) In fact, he confessed to one of his correspondents, J. H. Woodger (31/03/78),

© Springer Nature Switzerland AG 2022
F. Weinert, *Karl Popper*, Springer Biographies,
https://doi.org/10.1007/978-3-031-15424-9_1

that he was dissatisfied with all religions he knew because of 'their dogmatism and their intolerance' [568.03].[1] His religion was the magnificence of the world. (*All Life* 1999: 42) But he remained a social democrat at heart even after he had turned his back on communism. Popper's political outlook was also inspired by Kant's republicanism. There is unity in his thought, but it underwent revisions and shifts. In this sense his own work displays the evolutionary trend that was characteristic of his later work in the 1960s and 1970s.

In an interview in the 1980s, Popper claimed that by 1919 he had formulated his major problems. (*Offene Gesellschaft* [3]1983: 36)[2] This claim is not quite correct. What we find is both continuity and discontinuity. In his psychological work on learning processes, between 1921 and 1926, he came to realize the importance of inborn dispositions, i.e. of some form of 'inborn' knowledge. In the 1950s he changed his views on physics, from a deterministic to an indeterministic stance. In its wake, he developed a new account of probability: the propensity theory. He also revised his assessment of the scientific status of Darwin's theory of evolution. His early emphasis on falsification gave way to a trial-and-error method. Into his old age, he remained committed to cosmopolitanism, principles of democracy in an open society and the importance of peace. He shared these ideas with the German philosopher Immanuel Kant, who had a lifelong influence on his thought. Two problems dominated his first excursions into philosophy: the problem of demarcation of science from non-science and the problematic nature of inductive inferences from past to future events. But he did not realize until the late 1920s how closely related these problems were.

These shifts were connected with an increasing emphasis on the notion of time. Popper argued in favour of the Heraclitean view that the world is in flux. Change is fundamental. There is no mention of the importance of time in his *Logik der Forschung* (1935),[3] his first major book. But it became central to his thought, as did his later evolutionism. There is no mention of Darwin either, but there is, in retrospect, an important passage on the natural selection of competing theories (§30). In the 1960s Popper made a decisive turn towards evolutionary views of the growth of knowledge, in response to criticism. He proposed an improvement of evolutionary theory, which

---

[1] The rights to Popper's writings are held by the University of Klagenfurt, which has given permission to quote, directly or indirectly from his works: @2008 University of Klagenfurt/Karl Popper Library. All rights reserved. The originals are held at the Hoover Institute Archives, Karl Popper Papers. The catalogue is available at http://pdf.oac.cdlib.org/pdf/hoover/reg_189.pdf. The citation format adopted here is: [box number.file number].

[2] This book has been translated into several languages, but is unavailable in English. It is a record of his conversations with the Austrian journalist Franz Kreuzer, from 1979 to 1982. Kreuzer also edited the texts of the Vienna Popper Symposium under the title *Die Zukunft ist offen* (1985), which is equally unavailable in English. The German newspaper *Die Welt* conducted two interviews with him (1987 and 1990), which are published in *Ich weiß, daß ich nichts weiß* (1987/[2]1992).

[3] This book was published into English under the title: *The Logic of Scientific Discovery* (1959); a better translation would have been *The Logic of Scientific Inquiry*. The term 'logic' is an apposite reminder that the book is not intended as a description of scientific practice but as a prescription, a normative methodology, of how scientific research should be conducted.

in the 1980s led to a distinction between passive and active Darwinism. Darwinism inspired his attempted solution of the mind–body problem. Popper always regarded himself as a realist. But his realism grew more sophisticated, as he developed his interest in Greek cosmology and argued for a dynamic view of time.

In the 1970s and 1980s, as his philosophical influence amongst professional philosophers began to wane, he turned into a public intellectual, especially in Continental Europe. He also came to be appreciated in Iran, Turkey and Latin America. One of his most loyal admirers and supporters, Bryan Magee, reported an 'explosion of "Poppermania" in Italy', (27/01/83) [323.04].[4] But according to the correspondence Magee received at the time he was still little known in Canada and underrated in the US [323.01-02]. Undeterred, Popper continued his fight for rationality in an open and free society, commented on issues in political philosophy but also spoke about political problems of the day, such as overpopulation, the voting system, the danger of television, and just wars. He was no longer an uncompromising pacifist.

Popper's output reflects an astonishing breadth of interests. It ranges from the natural to the social sciences, from epistemology and logic to evolutionary theory, from biology to physics, from anti-authoritarianism to scepticism towards experts in all shapes and forms. Popper was no eclectic. He was a truly original thinker. He offered a new approach to the theory of scientific knowledge, a controversial interpretation of the history of political ideas, a new take on probability, an 'improvement' of Darwin's evolutionary theory; he opposed subjectivist interpretations of quantum mechanics, thermodynamics and relativity, the closed society and materialist solutions to the mind–body problem, one of the great remaining mysteries.

One enduring aspect of his career is the clarity of his style, which could mask the profundity of his thought. Popper hated verbosity. He preached and practised that even the most difficult and challenging ideas should be couched in simple language. Coupled with the clarity of his expression comes the precision of his thought. There is no triviality in his writings. He was not given to small talk, not even in his correspondence.

As a hobby joiner, I was immediately struck by his revelation (on page 1 of his *Autobiography*, 1974) that as a young man he spent an apprenticeship with a cabinet maker in Vienna (1922–24). His name was Adalbert Pösch. He claimed that he learnt more from Herr Pösch than from anyone else. (A few chapters later [1974: §17, p. 70] he said the same of the Polish logician Alfred Tarski.) He learnt from this craftsman the Socratic insight that he knew very little. It became a guiding principle of Popper's life. Herr Pösch, by contrast, boasted that he knew everything. He would not have taught Karl about the *Vienna Circle*, the dominant philosophical school at the time,

---

[4] The Turkish Foreign Ministry contacted Magee to ask whether he would be willing to discuss with the Minister of Education a textbook on Critical Thinking to be used in Colleges (09/11/78) [323.02]. Popper's wife, Hennie, proudly informed Magee (17/09/80) that 'Karl has become a foreign member of the oldest still existing academy: the Academica Nazionale dei Lincei (Rome). Galileo was a member' [323.03; underlined in original].

the history of philosophy, classical and modern physics or evolutionary theory. But Herr Pösch, the cabinet maker, would have inculcated in Karl a love for precision that he never lost. Attention to detail, precise measurement and care for the finished product are the hallmark of a good cabinet maker.

One of his biographers claimed that 'he was a bit of a hypochondriac'. (Watkins 1997b: 681) But his letters tell another story: throughout his life he suffered from poor health (bad teeth, bouts of pneumonia, headaches, lumbago, minor strokes, pleurisy, physical breakdowns, tachycardia and tinnitus). He was also allergic to smoke, as he reminded some of his correspondents: Alfred Ayer [270.35], Viktor Kraft [316.24] and Thomas Kuhn [317.17]. He frequently told his friends that he was depressed, yet he remained a workaholic all his life. He and his wife Hennie were often exhausted, overworked and short of money. In his seventies increasing deafness caused him enough discomfort to decline invitations. Visitors saw no television in his home. He informed them that he did not read newspapers. One of his best friends, the physicist and philosopher Wolfgang Yourgrau attested that he was 'a difficult person' (10/08/76) [364.12; my translation]. Some episodes reveal this difficult side of his personality. He suffered from what Magee characterized as 'neurotic perfectionism' [322.18; 359.33], coupled with frugal habits and a strong will [323.03]. But his voluminous correspondence also reveals a man of generosity and loyalty towards colleagues and friends. His letters contain no disparaging remarks or personal attacks even on his most severe critics. One of his former students, Joseph Agassi, described his personal characteristics as 'boundless dedication, stern moralism and a tendency to complain'. (Agassi 2008: Abstract) Some of his former assistants and colleagues (such as William Bartley, David Miller, Ian Jarvie, Arne Peterson, Jeremy Shearmur, Colin Simkin and John Watkins) remained on good terms with him, even through choppy waters. As we shall see, he kept up a cordial and supportive correspondence with many of the leading intellectuals of his day. When asked he was always willing to help colleagues to find academic positions, for instance William W. Bartley, Paul Feyerabend, Alfred Tarski and Wolfgang Yourgrau. In New Zealand he and his wife tried to help refugees from Nazi Germany. Like other people, he sometimes fell out with friends and colleagues, most notably Joseph Agassi, Imre Lakatos, even William Bartley or ill feelings erupted, as with John Watkins, at least for a while. This seemed to happen primarily when he had the impression that he had been misunderstood. Despite his many public awards and recognitions, he felt unappreciated by his contemporaries.[5]

In this book I shall endeavour to present his ideas, sympathetically but not uncritically. It is in the spirit of his own work to keep a critical distance. My emphasis will be on the key aspects of his life and ideas, rather than technical details of his work. My aim is to present Popper as a thinker whose views are still worthy of our

---

[5] For instance he complained in a letter to fellow Austrian Friedrich von Hayek (11/12/84) that he was constantly being attacked 'in the most dishonest way from all sides' [305.17].

attention. This is in agreement with Popper's own ambition who once told a correspondent—P. A. Schilpp (09/05/63)—that 'I have always seen it my duty to apply my epistemology to the most burning question of the day' [330.02]. I shall discuss what scientists think of his criterion of falsifiability and his contributions to biology and physics. The numerous publications that are still devoted to his work, as listed in the *Newsletters* of the Karl Popper Foundation in Klagenfurt, bear evidence to the ongoing topicality of his philosophy. He started out as a school teacher, turned to professional philosophy and finished his distinguished career as a public intellectual. He brought about a change of paradigm. One of his disciples, Imre Lakatos, regarded his ideas as revolutionary. But Popper fell out with him.

Chapter 2
Popper's Early Life in Vienna

I do not believe that the conjectural character of all our knowledge would ever have become clear to me without the work of Newton and Einstein….(K. Popper, *The World of Parmenides* 1998: 50)

2.1   In His Parents' Home

Karl Raimund Popper was born on July 28, 1902, at Himmelhof in Ober Saint Veit on the Western outskirts of Vienna, the capital of Austria.[1] His father, Simon Carl Siegmund Popper, came from Bohemia. His mother, née Jenny Schiff, was born in Vienna. They married on April 3, 1892, in Vienna's main synagogue. But in 1900 they both converted to Lutheranism in an effort to fit into the predominantly Christian culture of Austria. His mother came from a musical family and instilled in Popper a lifelong love of classical music. He even started composing at the age of sixteen. Only one composition, a fugue for organ, has survived. The family lived in a large apartment in the centre of Vienna, near the Stephan's Cathedral (*Stephansdom*), on the *Freisingergasse* 4 (which later became *Bauernmarkt* 1). The office of his father's law practice formed part of the apartment. Simon Popper, who had been a partner in the law firm of Raimund Grübl, Vienna's last liberal mayor, took over the practice after Grübl's death in 1898. (In his *Autobiography* Popper erroneously gives Grübl's first name as Carl.) He had two older sisters, Emilie Dorothea (Dora) and Anna Lydia (Annie). Dora committed suicide in 1932. Annie, whose married name was Lothringer, became a writer of romantic stories and would settle in Ascona,

---

[1] I have drawn primary information about Popper's life and times from his extensive and voluminous correspondence. The reader should be aware that it reflects how Popper interpreted events, not necessarily how they actually happened. There are missing or lost letters. For further information on his life I have used Bartley (1989); Brudny (2002); Geier (1994); Hacohen (2000: Ch. I); Miller (1997); Watkins (1997a, b) and Popper's *Autobiography* (1974). All life dates are given in the Index of Names.

© Springer Nature Switzerland AG 2022
F. Weinert, *Karl Popper*, Springer Biographies,
https://doi.org/10.1007/978-3-031-15424-9_2

Switzerland. Popper is silent about his sisters in his *Autobiography* (1974) but he writes affectionately about his parents.

Karl grew up in a musical, learned and pacifist household. His father was a successful lawyer, with wide-ranging intellectual interests. It was from his father that Karl first learnt about Darwin, as Popper told his long-term friend Konrad Lorenz (letter 21/01/75 [321.04]). His library contained a rich collection of books (12,000–14,000), including volumes by pacifist authors. Karl was particularly impressed by the Norwegian explorer Fridtjof Nansen, whose book *Farthest North* (1897) he read as a seven-year old. The book, which describes the Norwegian polar expedition (1893–1896), left a profound mark on the young Popper. He explained the influence much later on the occasion of receiving the Otto Hahn Peace Medal, in 1993. Nansen's 'voyage through three long Arctic winters became an experimental test of his daring but well-thought-out theories'. (*Alles Leben* 1996: 321–322; *All Life* 1999: 141) Popper did not mention Nansen in his *Autobiography* (1974) but he mentioned Bertha von Suttner, the first woman to be awarded the Nobel Peace Prize (1905). His father's library contained her book *Die Waffen Nieder!* (*Lay Down your Arms!*). First published in 1889, it became an international bestseller. Popper often talked about von Suttner and her influence on his thinking in his later life, according to Hubert Kiesewetter, a German economist and social historian, who knew Popper from his time at the *LSE*. Apart from Nansen and von Suttner, a third moral influence on the young Popper was Josef Popper-Lynkeus. He was an engineer and social reformer, a member of a left-leaning reformist circle called *The Monists*. (Kiesewetter 1995; Hacohen 2000: 43–60) These books, Popper said, were part of his life. (*Autobiography* 1974: 6) As we shall see, of equal, if not of greater importance, for his revolutionary ideas on science were Albert Einstein, Bertrand Russell and Immanuel Kant.

Bertha von Suttner (1843–1914), pictured in 1906. *Source* Wikimedia Commons

For the puritanical, even 'priggish' boy that he was (*Autobiography* 1974: 4) Karl's scholastic experience was not a happy affair. He attended a private elementary school, from 1908 to 1913, where the teaching was progressive. This was followed by attendance at the Realgymnasium, the Franz-Joseph Gymnasium in 1914 and a further Gymnasium for two more years. (Hacohen 2000: 65, 73–74) But he found the experience at these schools uninspiring. As he reported in his *Autobiography* he was appalled by the poor quality of the teaching. It was 'boring'. He left school and home in the winter of 1918–1919. He moved into a converted military hospital, which he shared with fellow students. (His father had lost his fortune in the inflation after World War I.) He enrolled at the University of Vienna, first as a non-matriculated student, and became a member of its Mathematical Institute. In 1921 his first attempt at the *Matura*—the entrance examination for the University—was unsuccessful because he failed in Latin and Logic. He obtained his *Matura* on his second attempt in 1922. (Miller 1997: 370; Hacohen 2000: 103) Popper passed his *Matura* as a private pupil.

> I now became a fully matriculated university student. A year later I passed a second *Matura* at a teachers' training college, which qualified me to teach in primary schools. I took this examination while learning to be a cabinetmaker. Later I added qualifications to teach mathematics, physics, and chemistry in secondary schools. However, there were no posts available for teachers, and after concluding my apprenticeship as a cabinetmaker I became, as I have mentioned, a social worker (....) with neglected children. (*Autobiography* 1974: 31)

Josef Popper-Lynkeus (1838–1921). *Source* Wikimedia Commons

In later life, on occasion of the Kyoto Prize for Creative Arts and Moral Sciences (1992), he explained 'why he had no prospect of obtaining a teaching position, not even in a primary school'.

> I had been too young by one year to fight as a soldier in the Imperial Austrian Army in the First World War; and all the available positions for teachers were, quite properly, reserved for the soldiers who came back from the war, or from the prison camps. (*All Life* 1999: 158)

Apart from his attendance at University, he tried his hand at manual work, under the spell of his socialist convictions. He quickly left a job in road construction as too exhausting and turned to social work. More enduring was his apprenticeship with the cabinet maker, the afore-mentioned Adalbert Pösch (from 1922 to 1924). He learnt enough from Herr Pösch to build some of his own furniture, later, in his house in Christchurch, New Zealand.

Arthur Stanley Eddington (1882–1944), British astronomer and philosopher of science. *Source* Wikimedia Commons

2.2 Intellectual Influences

It is often said that the year 1919 was of crucial importance in Popper's intellectual development. But it should not be forgotten that earlier events prepared the ground for Popper's later intellectual accomplishments. The year 1905 marked the birth of Einstein's Special theory of relativity[2] and the publication of Freud's *Three Essays on Sexuality*.[3] Popper later described how the examination of Freud's dream analysis provoked his views on demarcation: how to separate scientific from pseudo-scientific claims. (*Realism/Aim* 1983/1985: Part I, Ch. II §18, p. 174) His indebtedness to Einstein is repeated many times. After 1905, both Freud and Einstein continued to develop their respective theories. Freud arrived at his tripartite model of the mind (Id-Ego-Superego). In 1916 Einstein extended his Special to his General theory of relativity, from which he derived three quantitative predictions. Whilst the Special theory had dealt only with uniform motion, the General theory included accelerated motion. Three years later, in 1919, the English astronomer Arthur Eddington organized two expeditions, one to the island of Principe on the west coast of Africa and

---

[2] And his discovery of the law of the photoelectric effect, for which he received the 1921 Physics Nobel Prize. The photoelectric effect describes the emission of electrons when electromagnetic radiation, such as light, hits a metal surface. Electrons emitted in this manner are called photoelectrons.

[3] Freud's sister, Rosa Graf, was a family friend. She holidayed with the Poppers in Bad Ischl in 1916. She died in Treblinka in 1942.

one to Sobral in Brazil. They confirmed one of the predictions of the General theory. The success of Eddington's expedition turned Einstein into a celebrity overnight. The prediction was that light from distant stars would be deflected by a calculated amount (1.75") when it grazed a strong gravitational field, like that of the sun. This meant that a star's apparent location, as seen through a telescope from earth, was not its actual position. The line of sight is shifted by the gravitational effect of the sun on the star's light rays. (Fig. 2.1a) The photographs taken during a solar eclipse (when the moon blocks out most of the sunlight) were then compared to images of the stars' normal positions in the sky. The comparison of the images shows the apparent shift. (Fig. 2.1b)

Albert Einstein in Vienna, 1921. *Source* Wikimedia Commons

As mentioned, Popper would later declare, somewhat misleadingly, that by 1919 all his major problems were in place. Why was this year so important in his life? Firstly, there was the triumph of Einstein's theory thanks to the success of Eddington's expedition. Einstein visited the University of Vienna in the first part of January 1921 to give two lectures on relativity to invited guests. They were held at the *Chemisch-Physikalische Gesellschaft*, following the invitation of its director Felix Ehrenhaft.[4] A day later he also presented a popular lecture on relativity, to a packed hall, at the *Wiener Urania* (10. January 1921), as reported in the *Linzer Tages-Post* (11. January 1921). It was followed by a discussion the following day. The *Urania* in Vienna is still a public educational institute and an observatory. In his *Autobiography* Popper did not specify, which of the three lectures he attended. Perhaps it was the popular lecture at the *Urania*, since Popper was associated with the Mathematical Institute (Boltzmanngasse). It left him so 'dazed' that a fellow student, Max Elstein, had to explain Einstein's theory to him. Elstein, who died at the age of 21, was a brilliant student of mathematics. The encounter with Einstein's theory awoke Popper from his 'Kantian slumber'. It made him realize that well-established theories, such as

---

[4] Felix Ehrenhaft appreciated Popper as a clear thinker [292.07]. Popper knew Ehrenhaft as a 'brilliant experimentalist'. (See 'Replies' 1974: 1193 [fn 166])

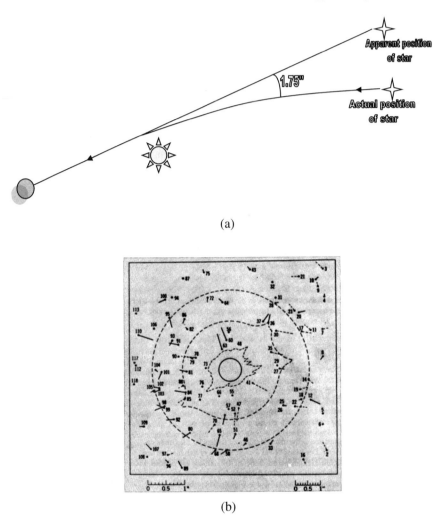

(a)

(b)

Fig. 2.1 **a** Bending of light by the gravitational field of the sun. From earth, the distant star is not seen in its actual location. According to Einstein's General theory of relativity, the deflection is a consequence of the curvature of space-time, due to the sun. **b** Image from 1922 of the actual measurement of the deflection of light. *Source* R./H. Sexl, *Weiße Zwerge, Schwarze Löcher* (Vieweg 1979: 14)

Newton's mechanics, can be replaced. There was no certainty in scientific knowledge. It set him on the road to his famous falsifiability criterion.

Karl at fifteen in 1917. *Source* Wikimedia Commons

Secondly in that year, he also found reason to doubt the 'scientific' status of psychology. He knew Alfred Adler, the founder of the school of individual psychology. He presented Adler with a case that did not seem to fit easily into his theory of individual psychology. But Adler had no difficulty in accommodating the case. He met in Adler the same willingness to dismiss seeming counterevidence, as in Freud. (See *C&R* 1963: 35).

Thirdly, a political event, also in 1919, overturned Karl's political sympathies. Popper had considered himself a communist, if only for a few months. In the spring of that year he had joined the Communist Party because he believed that it abhorred violence and fought for peace. (*Autobiography* 1974: 24; *Lesson* 1997: Interview I: 14–5) The young Popper, who had witnessed misery and poverty amongst the Viennese population, developed a passion for social justice. It never left him. Karl offered his services as an errand boy to the local branch of the Communist Party. Then a crucial event happened, which led him to an examination of his communist convictions. The event occurred in Vienna's *Hörlgasse* (9th district) on June 15, 1919. Karl witnessed how police shot some of his comrades during a peaceful communist demonstration. According to his recollections eight demonstrators were killed. (*Alles Leben* 1996: 309; *All Life* 1999: 133, 108; *Offene Gesellschaft* [3]1983: 7) Other sources, however, speak of twenty young communists killed and seventy severely injured demonstrators. (*Zukunft* [4]1990: Vorwort; cf. Geier 1994: 27) Popper was horrified to see how the party leadership presented the deaths of his young comrades as 'collateral damage'. It was the price to be paid for the proletarian struggle against the evils of capitalism. He often recalled the events in *Hörlgasse*. But on July 15, 1927, he and his future wife Hennie experienced another example of police brutality.

On that occasion, eighty-nine protesters and five policemen died in what is known as the July Revolt of 1927. (*Autobiography* 1974: 85)

When later interviewed about these events he described the realization to which they had led him. Communism taught that the revolution would come; it was inevitable. Nothing would stop it. Hence resisting this historical force was counter-revolutionary. The struggle also brought with it its necessary sacrifices. In a long, handwritten letter to the Dalai Lama (18/12/91) Popper admitted that

> (w)hen I was a Communist, 74 years ago, I believed that it was possible to 'prove scientifically' that communism (or socialism) is bound to come: I believed that Marx had proved it. (…) In believing this, I felt bound by moral duty to support the Communist party: the party whose historical task it was to help to bring about Communism [287.06].

But in the same letter he confessed that the 'close contact' with communist leaders and the events in Russia taught him that they were 'horrible liars', willing to commit murder. Marx had made a mistake: the future cannot be predicted. Marxism was 'not a science'.

The pretence that history is subject to 'iron laws of development' was the major criticism, which Popper levelled at 'false prophets' (Plato, Hegel and Marx). After the events in *Hörlgasse* Popper decided, in July 1919, to review his attitude towards Marxism. Here was another 'theory' which could not stand the test. In February 1920 he reached his final decision: he turned his back on Marxism but remained a social democrat. In an interview with the Italian journalist G Bosetti (published posthumously in *The Lesson of this Century*, 1997) he summarized the main argument in Marx's *Das Kapital*: **a)** capitalism cannot be reformed, it must be destroyed; **b)** capitalism is responsible for increasing misery; **c)** capitalists themselves are victims of the system. There was no point in tinkering with the system at the edges. It had to be overthrown. The Party demanded absolute loyalty to the cause. Popper felt caught in this trap before he started critically examining the central tenets of Marxism. What freed him from this 'prison' was the realization that history was not like an unstoppable river. It did not obey iron laws. Marx, of course, was not the only nineteenth-century thinker who believed in historical laws. He shared this belief with English philosopher John Stuart Mill and French philosopher Auguste Comte. So Popper's examination of Marxism had wider significance. The question was whether social systems are subject to social laws in the same way that physical systems obey physical laws. In the work he would produce in New Zealand he answered decisively 'no' and gave reasons for his negative assessment. After having escaped from the 'cage', he never looked back. But in 1935 he began to formulate his criticism of Marxism. He was committed to freedom and piecemeal social reform. His famous plea for an 'open society' was born. He also was a fervent believer in the value and power of education and rationality. It is no surprise that Popper, in retrospect, would claim that by 1919 all his major problems were in place.

He demonstrated his social conscience in various ways. In the mid-20s, he worked for a while as a social worker and recalled how he sheltered a woman from her husband's aggression. (*Lesson* 1997: 59) In New Zealand he tried to help refugees from Nazi Germany. All his life he stood up for social justice. He also felt the lure of philosophy. He read Kierkegaard and tried his hand at Schopenhauer and Kant. He did not regard Kierkegaard's musings as philosophy, as he wrote to Magee (02/01/69) [322.18]. But he praised Schopenhauer as an 'honest thinker' (letter to Gombrich 23/08/43 [300.02]) and 'a genius' (letter to Magee 09/01/74 [322.20]). In the 1980s he would continue his discussions about Schopenhauer with Magee [549.19]. Kant, however, remained an enduring influence. 'I also started fighting my way through the *Critique of Pure Reason* and the *Prolegomena*'. Popper became a Kantian 'of sorts'. But Kantianism needed to be made compatible with Einstein's discoveries. Kant had taken the truth of Newton's theory for granted, whilst Einstein had questioned it. The solution appeared in his *Logik*: 'falsificationism' or better 'fallibilism'.

As a university student, Popper was not thrilled by the lectures. Only the classes on science and mathematics captivated him. But mathematics and physics are difficult subjects. Popper did not see his future as a mathematician. Rather he dreamt of starting a school, which would teach young people how to pose and solve problems. Learning without boredom. This concern for problems and attempted solutions became a dominant theme in Popper's philosophical career. He later called it 'the logic of the problem situation' (or 'situational logic').

A year before Popper joined the *Pedagogic Institute*, he met academics who had some impact on his thinking. In 1924 he made the acquaintance of the economist and social theorist Karl Polanyi, with whom he held discussions about the difference between the natural and the social sciences. Popper liked him but in retrospect complained to the art historian Ernst Gombrich (04/12/43) that he had 'failed him badly' in finding a publisher for his first book *Die beiden Grundprobleme der Erkenntnistheorie* [300.02].[5] Karl had a brother, Michael Polanyi, whom Popper would, thirty-five years later, regard as a 'post-rationalist'. That is, he would disagree with Michael's emphasis on the role of 'personal knowledge' in the natural sciences. In that year he also met philosopher Julius Kraft who tried to help Popper find a publisher for *Grundprobleme* (which became *Logik der Forschung*). Julius and Karl had night-long debates about Marxism and Kantian philosophy. Julius was a non-Marxist socialist, which made it easy for them to agree on Karl's critique of Marxism. They also shared the rejection of the philosophy of Edmund Husserl and Martin Heidegger. But they disagreed about Kantian epistemology, especially the role Kant's *synthetic* a priori propositions. Popper could not accept them as 'true' or 'proven'. But he was sympathetic to Julius's naturalistic view of the human or social sciences, as defended in his book *Die Unmöglichkeit der Geisteswissenschaft* (1934) (*The Impossibility of the Human Science*) [316.23].[6] (Hacohen 2002: 121–2, 358; Watkins 1997a: 211) They became lifelong friends. At that time Popper still believed that his book - 'an unusual achievement' - would secure him an academic position, as he wrote to Julius in an undated letter from the early 1930s [316.23; Hansen 3.2]. (Prior to its publication he seems not to have envisaged an academic career, as he

---

[5] This book was finally published in German in 1979, edited by Troels Eggers Hansen, and translated into English under the title *The Two Fundamental Problems of the Theory of Knowledge* in 2009.

[6] In his Obituary of Julius Kraft, Popper (1962) curiously translates the title of Kraft's book as The *Impossibility of a Spiritual Science*.

told Konrad Lorenz in January 1975 [321.04]). The two men lost sight of each other when Kraft emigrated to the USA in 1934. But in 1954 he came to London on a research fellowship and began to attend Popper's famous Tuesday seminars at the *LSE*. He launched the philosophical journal *Ratio*, which is still influential today. In 1957 he returned to Frankfurt.

2.3   At the Pedagogic Institute

In 1925 the city of Vienna founded the *Pedagogic Institute*. The authorities made it compulsory for all aspiring teachers to attend the Institute. Popper had left the apprenticeship with Herr Pösch and was working with neglected children. He applied and was admitted to the Institute where he had to take lectures on psychology. The remit of the Institute was to support the reform of primary and secondary education in Vienna. Given Popper's interest in education and learning he was not out of place. The best-known teacher at the Institute was the gestalt psychologist Karl Bühler. Popper attended his lectures and seminars and read his books. He stayed in 'loose touch with Karl Bühler from 1924 to 1936' but took an 'unjustified dislike' to his wife, as he admitted in a letter to Lorenz (21/01/75; my translation). This led him to less frequent attendances at Bühler's seminars but did not stop him from successfully tutoring dozens of Bühler's best students [321.04].

Bühler saw the child as an active social being, capable of intellectual and critical judgement. It laid the foundation of Popper's lifelong opposition to what he called 'the bucket theory' of the mind. It describes the empiricist view that the mind is like an empty slate, on which our growing experience of the world begins to write. It is the idea that our senses fill our mental container with data. Bühler's theory of language also influenced Popper's contributions to the philosophy of mind: his explanation of the emergence of mental processes. Bühler distinguished the expressive function, the signalling function and the descriptive function, to which Popper would later add the argumentative function. (The expressive function allows a speaker to express emotions and thoughts; the signalling function allows the listener to react to the speaker; the descriptive function allows the speaker to describe states of affairs; the argumentative function is the basis of critical thinking because its focus is on arguments. See *C&R* 1963: 134–5)

Popper also made the acquaintance of a fellow student, Anna Josephine Henninger. Her nickname was Hennie. They married in 1930. The couple decided not to have children. She was to become a source of inspiration for Popper. She would retype his manuscripts numerous times and became, by Popper's own admission, 'one of the severest judges of my work'. (*Autobiography* 1974: 57) He dedicated the English translation of *Logik* 'to my wife who is responsible for the revival of this book'.

In his second year at the Institute Popper met the Austrian philosopher and Greek scholar Heinrich Gomperz. He cited him in his first (then unpublished) book *Die beiden Grundprobleme der Erkenntnistheorie*. He had many discussions with him about the 'psychology of knowledge'. Gomperz distinguished psychology from logic. It was the beginning of Popper's lifelong anti-psychologism, which is an outstanding feature of his teaching and writing. As he explained in his *Autobiography* (1974: 59): 'I was reacting more and more strongly against any "psychologistic" approach, including the psychologism of Gomperz'. In a letter, dated February 28, 1935 (the name of the recipient is left out, perhaps it was Gomperz) he characterized psychologism as the view that verdicts about perceptions are of epistemological interest [017.06]. Later he would praise Marx for his anti-psychologism.

Popper became convinced that psychological approaches, using sense data or impressions (as the British empiricists held), cannot form the basis of a theory of (scientific) knowledge. The mind is not a receptacle for collecting and storing information about the external world. It is a creator of daring hypotheses and universal theories. He adopted a position of objectivism. It emphasizes the 'priority of the study of logic over the study of subjective thought'. (*Autobiography* 1974: 61)

At the Institute he submitted, in 1927, an unfinished thesis on 'habits and experience of laws in education'. This piece was followed, in 1928, by his PhD thesis under the title: *On the Problem of Method in the Psychology of Thinking.* It marked his transition from psychology to logic, from subjectivism to objectivism. In his viva (*Rigorosum*) he was examined by Karl Bühler and Moritz Schlick. Schlick was the leader of the influential Vienna School of philosophy, the *Vienna Circle.* Bühler allowed Popper to talk about the logic of science. But Schlick examined him on the history of philosophy, especially Leibniz's views on idealism and realism. Popper felt that his answers were not to Schlick's satisfaction. His own impression was that he had not done very well. To his surprise he was awarded the highest available mark and passed with distinction. (*Autobiography* 1974: 62; [252.01]) In order to qualify as a teacher in secondary schools, he submitted, in 1929, a third thesis on 'axioms, definitions and geometry'.

Moritz Schlick (1882-1936), philosopher, physicist and logical positivist. *Source* Wikimedia Commons

In the second part of the 1920s, he wrote several pedagogical articles on school reform, pioneered by the city of Vienna. They stressed the importance of treating pupils as individuals and recommended their active participation in the learning experience, as against the practice of rote learning. The essays reveal an early hint of Popper's adoption of individualism as a methodology in the social sciences. Also notable is his article 'Zur Philosophie des Heimatgedankens' (1927). In this piece on the philosophy of the notion of *Heimat* (homeland), Karl revealed his antipathy to nationalism and contrasted it with a 'cosmopolitan socialist vision for progressive education'. (Hacohen 2000: 127) This vision later morphed into his Kantian cosmopolitanism.

Now Mr. Popper had become Dr. Popper but his Jewish ancestry made it practically impossible to obtain a university position. Anti-Semitism was widespread in Vienna and Europe at that time. Whilst Liberalism was in decline, nationalism was on the rise. (See Schorske 1981; Hacohen 2000: Ch. I)

In 1932 (June 22) Popper lost his father, for whom he felt much love and admiration. He was upset that he had to sell the large library of his father—'a true monument to his personality and his great and universal interests'—for 'ridiculous amounts of money' (undated letter to J. Kraft, probably from 1932 [Hansen 3.2]; my translation). His mother, of whom he was also very fond, died in May 1938.

2.4   The Vienna Circle and Visits to England

Popper began to associate with *The Vienna Circle* in 1926–27. He would have liked to become a member, as he admitted in an interview in the 1980s, but Schlick never invited him. He regarded himself as one of his 'pupils' and admired Schlick's *Allgemeine Erkenntnistheorie* (1918, [2]1925), for its defence of realism and criticism of positivism. This 'immortal book', as he later called it (letter to Feigl 02/11/74 [536.10]) was translated into English under the title *General Theory of Knowledge* (1974). (Popper was very pleased with Watkins's review of the book (25/04/77) [360.02]). However, when Popper met Schlick for the first time in 1928 he had abandoned his realism, under Wittgenstein's influence. Popper told him that it was for him a 'personal catastrophe' [252.01]. He was also critical of Schlick's physicalism, his Wittgensteinian views on causality, natural laws and his sympathy with Heisenberg's interpretation of quantum mechanics. Herbert Feigl, another member of the Circle, also admired Schlick and wrote the introduction to the English translation of *Allgemeine Erkenntnistheorie*. But he was also supportive of Popper's work, not only when they met in 1930 but throughout their lifelong friendship. Feigl emigrated to the USA and finished his career as Director of the Centre for Philosophy of Science at the University of Minnesota. Popper's encounter with the philosophical ideas of the *Vienna Circle* ultimately led to the publication of *Logik der Forschung* (1935).

The philosophers of the *Vienna Circle* had sought to find a way to distinguish real science from metaphysics. They adopted the *verification* principle as a demarcation criterion of both scientific and meaningful statements. For a statement to count as meaningful it had to be empirically verifiable. Its meaning was bestowed on it by scientific research. In the eyes of the *Vienna Circle* science dealt with facts and observations, not speculations. The facts would accumulate until—by inductive generalization—a more general hypothesis or theory could be extracted. The facts would confirm or verify the theory. Meaningful empirical statements could be retraced to their empirical origin. Thus the method of induction—the extrapolation from particular to more general and universal statements—became their demarcation criterion. This approach would show, according to the logical empiricists, as they were also called, that metaphysical statements were meaningless. They took their inspiration from Wittgenstein's *Tractatus* (1921), which asserted that 'whereof you cannot speak, thereof you must be silent'. Metaphysical speculations certainly had no place in science. In *Language, Truth and Logic* (1936), the Oxford philosopher Alfred J. Ayer presented logical positivism to the English-speaking world. Its first chapter dealt with the elimination of metaphysics. One way of achieving this was the adoption of the *verification principle*. This principle, which requires the truth or falsity of meaningful factual statements to be established by empirical means, had been introduced by Fritz Waismann (1930), another member of the Circle.

Popper's ideas were largely seen as a challenge to logical positivism, the official philosophy of the *Vienna Circle*. Popper did not agree with the dismissal of metaphysical statements. But at that time he was uncertain of how to deal with metaphysics. Only later did he come to realize that even if metaphysical systems cannot be refuted, they can at least be criticized. In his later work Popper found a more positive role for metaphysics even at the heart of science. Metaphysical convictions—for instance, that all matter consists of atoms, or that the universe is like a gigantic deterministic clockwork—could actually inspire scientific research.

In his *Autobiography* Popper presented himself as the slayer of logical positivism. This claim is debatable. Agassi (2008: Appendix 1) recalls that Russell had already criticized the logical positivists' commitment to the verification principle. In his *Problems of Philosophy* (1912) Russell discussed the knowledge of universals, of general principles and a priori statements. Knowledge of such abstract entities exceeded the requirement that the sense of a proposition consisted in its method of verification. There were other striking differences between the *Vienna Circle* and Popper: Popper did not believe that science could operate without metaphysical convictions and he replaced the principle of verifiability by his principle of falsifiability. (*Offene Gesellschaft* [3] 1983: 37–8)

The British mathematician and historian Jacob Bronowski described a feeling of dissatisfaction with scientific philosophy in England at the time.

> In 1930 the model of the philosophical method in Cambridge was still the *Principia Mathematica* by Alfred North Whitehead and Bertrand Russell, and the *Tractatus Logico-Philosophicus* by Ludwig Wittgenstein. (…) The final task of philosophy in science would be to establish a universal system of axioms from which all phenomena of nature could be derived.

He credited Popper with having helped to bring about a change of outlook:

What Karl Popper had to say was very timely, because it came when the climate was changing, and it helped to change it; we were conscious of that at the time, in philosophy as well as in politics. (Bronowski 1974: 606-7)

Young scientists at Cambridge were hungry for new ideas and a fresh look at science. Popper's book provided a fresh impetus. Although *Logik* was not translated into English until 1959, Popper's ideas were circulating in English intellectual circles. In fact, Popper was known well enough to earn himself an invitation to England.

But before coming to England Popper met Alfred Tarski, who will forever be associated with the correspondence theory of truth (see Box 2, Chap. 3.6, p. 49). Their encounters occurred:

...first at the Prague conference in August, 1934, when I had with me the page proofs of *Logik der Forschung*; in Vienna in 1934-35; and again at the Congress in Paris in September, 1935. And from Tarski I learned more, I think, than from anybody else. (*Autobiography* 1974: 70)

It proved to have a major impact on Popper's philosophy. He no longer had any qualms about speaking of truth and falsity. In his *Logik* (§84) he had avoided a philosophical discussion of the murky notion of 'truth'. He preferred to speak of deductive relations rather than truth or falsity. After several meetings with Tarski in 1934–35 he embraced the notion of truth without further reservations. Curiously, though, he had characterized truth and falsity in *Grundprobleme* (1979: 41; 2009: 37; italics in original) in terms of correspondence:

Every empirical statement can thus be regarded as the representation of a state of affairs. If the state of affairs represented by a statement actually exists (if, in fact, *there* is such a state of affairs), then the statement is *true*; if the state of affairs represented does not exist, then the statement is *false*.

However, Tarski had formally rehabilitated the correspondence theory of truth. That is, Tarski had found a logical solution to the problem of how true statements can be said to relate to facts, without getting involved in the liar paradox ('This sentence is false'). On a personal note, Popper admired Tarski and regarded him as a friend. The respect was mutual. They kept up a lifelong correspondence. Tarski, who wrote in flawless German, complained to Popper about the intolerable political situation in Poland. Already in 1936 he feared to lose his position as lecturer in the philosophy of mathematics at the University of Warsaw, as he saw his 'non-Aryan' colleagues forced to resign from their positions. He told Popper that he had no hope of securing a professorship. Nor did Popper. But he was willing to help Popper by writing a reference for him. But Popper was to write it himself so that he could simply sign it. The (undated) reference praised the author of *Logik der Forschung* as a 'thinker of great originality and profound knowledge' [406.05]. When he was in New Zealand Popper tried, unsuccessfully, to secure Tarski a position as 'statistician'. At least he helped him by looking over the translation into German of Tarski's *Introduction to Mathematical Logic* (1937). In 1939, when Tarski was on a lecture tour of the USA,—he was said to be an excellent lecturer [330.15]—the outbreak of war prevented him from returning to Poland. He joined the Department of Mathematics at Berkeley in 1942. The correspondence between the two men continued. In 1964 Tarski tentatively promised Popper a contribution to the forthcoming Schilpp Volume of Living Philosophers on Popper's philosophy. But this promise never materialized. Nevertheless, Popper dedicated his 1972 collection of essays *Objective Knowledge* to Tarski.

Alfred Tarski (1901–1983), Popper's inspiration. *Source* Wikimedia Commons

Popper's two visits to England took place in 1935–36, at the invitation of the British analytic philosopher Susan Stebbing. In January 1934, Stebbing had expressed an interest in having *Grundprobleme* translated into English. She had found a translator but Popper informed her that Volume II was not yet ready and Volume I still needed corrections (03/02/34) [352.15]. At London's Bedford College Popper gave his first two lectures in English, not on falsifiability but on Tarski's theory of truth. He also delivered three public lectures on probability in the Mathematics Department of Imperial College [Hansen 3.2], as well as two lectures in Cambridge and one in Oxford. He met the philosophers Alfred ('Freddie') Ayer, Gilbert Ryle, George E. Moore, Bertrand Russell and the influential physicist Erwin Schrödinger, who became a lifelong friend. They knew that Popper was looking for a position. In a letter, dated October 12, 1936, Ayer offered to introduce Popper to the Harvard philosopher and logician Willard Van Orman Quine with a view to securing him a job in America [270.35]. He also introduced Popper to the historian of ideas and political theorist Isaiah Berlin, with whom he would have a correspondence about the latter's two concepts of liberty. Popper retained a lifelong admiration for Russell. He came to regard him as 'perhaps the greatest philosopher since Kant'. (*Autobiography* 1974: 87) And Russell told Yourgrau (letter to Popper 10/12/76) that he regarded Popper 'as the best mind in philosophy of science in the world' [364.12]. The invitation to read a paper on 'The Poverty of Historicism' in von Hayek's seminar at the *LSE* would prove to be a life changer. Ten years later, Friedrich von Hayek, the Austrian-British economist, was instrumental in Popper's appointment to a Readership at the *LSE*.

Popper also met with a group of progressive molecular biologists who had formed the Theoretical Biology Club. He participated in their 1936 meeting. At their Oxford meeting of 1946 (July 26–29) he met the British biologist Peter Medawar who became a lifelong friend and proponent of Popper's 'falsificationism'. In that year he also became a member of the Club. (See Niemann 2014: Ch. I)

Chapter 3
The Discovery of the Falsifiability Principle

> Only if we are ready to accept refutations do we speak about reality. (K. Popper, *Conjectures and Refutations* 1963: 212)

Popper is most famous for his principle of *falsifiability*. It is striking that, throughout his career, he used three terms synonymously: *falsifiability*, *refutability* and *testability*. In order to appreciate the importance of these criteria it is helpful to understand (a) how he arrived at these notions, then (b) whether the conflation of these three terms is justified, even by the logic of his own system. And, looking at what problem(s) Popper tried to solve, the question arises (c) whether the principle works in real science. Why are some scientists unhappy with the invocation of falsifiability?

3.1 Towards a Logic of Research

Some of the intellectual influences on the young Popper have already been discussed. But how did he arrive at his notion of falsifiability? In order to explain its appearance at the beginning of his book *Logik der Forschung*, a bit of cultural history is needed. It is centred on Vienna. At the beginning of the twentieth century the Austrian capital was culturally a very progressive city. (Janik/Toumin 1973) It was home to the philosopher Ludwig Wittgenstein, the painters Gustav Klimt and Egon Schiele, the writer and satirist Karl Kraus, the composer Arnold Schönberg, the architect Otto Wagner and psychologists Alfred Adler and Sigmund Freud. Famous physicists such as Ludwig Boltzmann, Paul Ehrenfest, Joseph Loschmidt and Josef Stefan worked in Vienna. Loschmidt and Stefan were amongst the founding members of the aforementioned *Chemisch-Physikalische Gesellschaft*, where Einstein gave his lectures in

© Springer Nature Switzerland AG 2022
F. Weinert, *Karl Popper*, Springer Biographies,
https://doi.org/10.1007/978-3-031-15424-9_3

1921. It was also the home of the *Vienna Circle*—the group of like-minded philosophers who shared a keen interest in science and an equally keen abhorrence of metaphysics. Famous members were Rudolf Carnap, Otto Neurath, the afore-mentioned Mortiz Schlick and Viktor Kraft. The physicist Philipp Frank, the probability theorist Richard von Mises and the mathematician Hans Hahn also belonged to the Circle. The young Popper attended Hahn's lectures which he compared to works of art. Carnap became a notable philosopher of science in his own right. Popper acknowledged Carnap's importance when, in the Obituary he wrote for *The Times* newspaper (17/09/70), he called Carnap 'one of the most influential philosophers of our time, the undisputed leader of a school which called itself "logical positivism" and later "logical empiricism" (…)' [355.06]. In his later career, when Carnap worked on his theory of induction and logic of confirmation, he found himself at odds with Popper. Popper, for reasons to be discussed, preferred the notion of corroboration. Although Popper never became a member of the *Vienna Circle* he came into contact with some of its members. Viktor Kraft was his first contact. Kraft, like other members of the *Vienna Circle*, tended to downplay the differences between logical positivism and Popper's philosophy: both agreed on empiricism and emphasized the importance of testability. Both were committed to the Enlightenment, as Popper himself admitted. (*Autobiography* 1974: §17; Kraft 1974; [316.24]) Kraft published widely on the theory of knowledge. He came to see Popper as 'the sole competent judge' of his work [316.24]. Popper appreciated his book *Die Grundformen der wissenschaftlichen Methoden* (1925) ('The Basic Forms of Scientific Method') as a valuable contribution to epistemology (the theory of knowledge). Kraft adhered to 'scientific philosophy'. They shared the view that science was a deductive system. Kraft praised Popper's work as of 'outstanding importance' [undated review, Hansen 3.2].

Herbert Feigl found Popper's ideas 'almost revolutionary'. (*Autobiography* 1974: 65) He assured Gomperz (27/12/32) that Popper's manuscript contained the only theory of knowledge that could be taken seriously [Hansen 3.2]. It was due to Feigl's encouragement, and altogether to his credit, that Popper wrote *Die beiden Grundprobleme*.[1] The title of the manuscript was inspired by Schopenhauer's *The Two Fundamental Problems of Ethics* (1841). Popper never forgot Feigl's support.

> I want to thank you for your continued friendship which means much to me. As you know, far back in what seems like another life, you did me a great service in being prepared to listen to me, to argue with me, and to urge me to publish my ideas. (…) Without the advice you gave me far back in the early thirties I may never have published those ideas or, as I did late in life, I may have published them much too late. (Letters to Feigl 01/12/72 [536.10] and 23/03/60 [294.06])

---

[1] In a letter (30/11/32) to the publisher Paul Buske, Popper mentioned that J. Kraft had proposed two alternative titles: either 'The Philosophical Preconditions of Natural Science' or 'The Problem of Natural Laws' [Hansen 3.2; my translation]. Buske was one of the publishers on whom Popper pinned his hopes. Hacohen (2000): Chap. 6 provides a detailed account of the tortuous path of Popper's manuscript to its publication as *Logik der Forschung*. See also *Autobiography* (1974): 67.

Popper was aware that it could be an 'epoch-making book'. He shared with his friend Bertold (Berti) Wiesner, an Austrian biologist (22/05/32), his hunch that it probably counted amongst the 'most radical of its kind' and 'richest in ideas (and the least boring)' [362.08]. He circulated a few copies: to Feigl in America, to Michael Polanyi in Berlin, to Egon Friedell in Vienna and to Joseph Needham in England. He had spent seven years on the manuscript but he felt pessimistic about its publication chances [Hansen 3.2]. Before the book was even published, he had to deal with a priority dispute and an accusation of plagiarism, from an 'unassailable, famous name' (letter to Kraft, probably from 1932 [Hansen 3.2]). The two fundamental problems, the manuscript addressed, were Hume's problem of induction and Kant's problem of demarcation. The Scottish philosopher David Hume had expressed doubts regarding the justifiability of inferences from past to future events, from past causes to future effects. Immanuel Kant had sought to draw a line between scientific thinking and metaphysical speculation. Popper's original manuscript was far too lengthy and unwieldy (550 typed pages). But it captured the core of his philosophy. It was the work of a brilliant young philosopher who called for nothing less than a 'scientific reconstruction of philosophy'. (*Grundprobleme* 1979: 416, 2009: 380) Although Philipp Frank and Moritz Schlick, the editors of *Schriften zur Wissenschaftlichen Weltauffassung,* accepted the manuscript for publication as Volume 9, it required several cycles of rewriting. Additional efforts of Carnap (who was enthusiastic about *Open Society*), Gomperz[2] and a letter of recommendation from Schlick were needed before Springer offered Popper a contract. Popper was nervous about Schlick's support for he had launched a fierce attack on Schlick's idol ('Abgott') Wittgenstein. He asked Carnap to plead with Schlick for he considered that his 'book was impressive' (16/03/33; my translation). After a nervous wait, Schlick gave the manuscript a more positive evaluation than Popper had anticipated. He even agreed the title *Logik der Forschung* with Schlick [017.06; Hansen 3.2]. But after Popper had signed the contract, Springer demanded major changes to the manuscript. Walter Schiff, his uncle, made the final cuts.[3] Popper breathed a sigh of relief, as he confessed to Julius Kraft (11/07/33), when he held the contract in his hands. However, he regretted that radical cuts had to be made and that the book would appear in a collection that had such a 'positivistic orientation' [316.23; my translation]. His book was wrong for this collection [294.06]. And it had harmed his relationship with Schlick [017.06].[4] Popper saw in Stebbing's inquiry about an English translation another chance to get *Grundprobleme* published in full. The many cuts meant that it had practically become a new book [352.15]. Still it was *Logik* that Springer published in Vienna in 1935 and sold just 200

---

[2] Gomperz realized that Popper's book criticized the Vienna Circle, as he wrote to Popper (27/12/32). In a reference letter (21/12/32) to the publisher Paul Siebeck (of J. C. B. Mohr), Gomperz praised Popper's book for propounding, in clear language, a 'methodology of scientific knowledge', which remained close to the 'procedure of the mathematical natural sciences' and differed essentially from that of the Vienna Circle [Hansen 3.2; my translation].

[3] Walter Schiff, Popper's maternal uncle, taught economics and statistics at the University of Vienna.

[4] Schlick was murdered by a former student on 22 June, 1936, as he was leaving the university. In an undated handwritten note 'In Honour of Moritz Schlick' Popper conveyed the general impression at the time that he had been murdered by a Nazi [252.01], which is probably true.

copies in the first part of the year [316.33]. It was generally well received, for instance, by Carnap, but Frank and Neurath did not rate it highly. (Hacohen 2000: 275–289; Watkins 1997b: 649–651) The influential physicist and philosopher Hans Reichenbach, who was dismissed by the Nazis from his assistant professorship in the physics department of the University of Berlin, published a negative review in the journal *Erkenntnis* V (14/05/35). Reichenbach, a proponent of logical empiricism, judged the results of Popper's book to be 'untenable', not only because the formation of scientific hypotheses required inductive or probabilistic logic, but also because strict falsification was impossible [Hansen 3.2]. That is, Reichenbach made the mistake of treating Popper as a naïve falsificationist. Einstein, however, who received a copy of the book, was favourably impressed by its rejection of inductivism in favour of 'fallibilism'. It was not until 1984 that Popper became aware—when John Stachel[5] informed him—that Einstein had published a brief article in the *Berliner Tageblatt* (25/12/1919) entitled 'Induktion und Deduktion in der Physik' [292.12]. He argued that deduction had played a much more important role in the history of physics than induction. Einstein added that it was never possible to establish the truth of a scientific theory. It is no surprise that Einstein welcomed Popper's book.

Cover of 1st German edition of the *The Logic of Scientific Discovery* (Springer 1935)

Nor did Popper realize, in 1919, the close connection between the invalidity of induction and his principle of falsifiability (conceived as a demarcation principle). The connection occurred to him only ten years later as he told an audience in Cambridge in 1953 (*C&R* 1963: 39, 42) and repeated in letters to William Bartley (20/07/65) [272.02; cf. 271.22] and John Stachel (15/03/84) [292.12].

---

[5] In 1977, Stachel became the first editor of the Einstein Papers Project, then based at Boston University.

Psychoanalyst Sigmund Freud (1856–1939), pictured in 1905. *Source* Wikimedia Commons

Although Popper had understood very little of Einstein's Vienna lecture (1921), it left a deep impression on him. He detected in Einstein's theory a threat to classical physics. The replacement of Newton's long-established theory of gravitation—by the idea of curved space–time—helped Popper, so he said, to arrive at his conjectural theory of knowledge. (*Parmenides* 1998: 50) That is, all knowledge became fallible. This new view of scientific knowledge was a consequence of his criterion of falsifiability. Popper arrived at it because he was struck by the differences between Freud's and Einstein's theories. Throughout his career, Freud had consistently claimed that psychoanalysis was an exact science, in fact that it was on a par with physics. [6] Freud believed in the determination of psychic life, in which he found nothing trifling or arbitrary. Psychoanalysis could claim to be scientific because it was based on (as yet) undiscovered psychological laws; and because its fundamental tenet—the existence of the Unconscious (*Id*) and its dynamic influence on human behaviour—was supported by evidence. What he regarded as his body of 'firm evidence' boiled down to Freudian slips, neurotic behaviour and the interpretation of dreams. He famously called dreams the royal road to the Unconscious. But Popper remained unconvinced. He did not think that Freud was a charlatan—as he would later think of Hegel—but he saw that Freud was too uncritical towards his own theory, just like Adler.

For instance, Freud considered that his psychoanalytic theory of slips of the tongue was correct, *without* examining alternative explanations. (I use this example but Popper makes this accusation against Freud's dream analysis; see *Realism/Aim* 1983/1985: Pt. Chap. II, §18.) An alternative explanation of slips of the tongue is *spoonerism* or a linguistic transposition of letters or words. The effect is named after the English clergyman and Oxford scholar William A. Spooner, who had a reputation of inadvertently reversing the sound of two words, often to humorous effect. It is an everyday occurrence. In December 2010, the BBC news presenter James

---

[6] See, for instance, his *Outline of Psychoanalysis* (1938) and my discussion in *Copernicus, Darwin and Freud* (2009: Chap. 3).

Naughtie suffered an embarrassing slip of the tongue when he introduced the British Conservative politician and then Culture Secretary Jeremy Hunt as Jeremy C**t. Journalists were quick to dismiss this slip as an instance of spoonerism, jumping instead to a Freudian interpretation. It showed, they said, Naughtie's dislike of Mr. Hunt and the true feelings of journalists for politicians. Freud considered that the evidence he marshalled—however meagre it was—*confirmed* his theory. There was a strong tendency in Freud either to ignore contrary evidence or explain it away. Popper would no doubt have agreed with Einstein who wrote in a letter (1949) that Freud was overconfident in his own theory: '...no illusion lulled him asleep except for an often exaggerated faith in his own ideas'. (Quoted in Pais 1982: 515) That is, Freud sought verifications, not falsifications of his ideas. Einstein's theory of general relativity, by contrast, had made three numerically precise predictions, of which the deflection of starlight near strong gravitational bodies, like the sun, was one (Fig. 2.1a).[7] In Popper's view, such precise predictions had the advantage of making a theory easily testable. The more precise they were the easier they were to falsify. He therefore held that scientific laws or theories *exclude* numerous possibilities. If, for instance, the sun's deflection of starlight is 1.75 arc seconds, according to the General theory of relativity, then any other values, outside an error margin, are excluded. Unlike the loose, imprecise 'predictions' of a horoscope, scientific theories often commit themselves to quantitative precision.

The combination of all these elements—his scepticism towards the meaning criterion of the logical positivists, their insistence on verification and induction, the shaky basis of Freud's psychoanalysis, and the quantitative predictions of Einstein's General theory—prompted Popper to propose his falsifiability criterion. It was not meant as a meaning but rather as a *demarcation* criterion. It was to provide a better separation of genuine science from pseudo-science. The verification principle was both too narrow—it excluded metaphysical assumptions in science—and too broad—it admitted dubious 'theories' like psychoanalysis and astrology. Popper doubted both that all scientific thinking starts from basic observation statements, cast in physicalistic language, and that it generates universal knowledge by way of induction. He therefore disagreed with the logical empiricists on these fundamental points.

It follows that there is an asymmetry between verification and falsification. Scientific theories or laws are universal in scope. They cover both observed and unobserved cases, as well as actual and possible cases. ('If there were a tenth planet it would obey Newton's laws'.) It is therefore impossible to exhaust the evidence to make sure that all cases have been covered. Even if diligent observers travelled the world in an effort to confirm that all ravens are black, they could never be sure that all ravens had been observed. What if some clever raven hides from the approaching researchers? However, they can be sure that the statement 'All ravens are black' is

---

[7] The others were the perihelion advance of Mercury and the redshift of light in gravitational fields. In 1964, Irwin I. Shapiro proposed a fourth classic test: the time delay of electromagnetic radiation (such as radar signals) passing the Sun. Gravitational fields also have an effect on the ticking of clocks: a clock in a weak gravitational field runs faster than a clock in a strong gravitational field. In recent years, satellite-based tests have 'confirmed' (or in Popper's terminology, 'corroborated') the results of the classic tests.

false, if white ravens are found. Clearly the universal statement that 'all ravens are black' cannot be true if there exists a colony of white ravens. The example is not far-fetched. For a long time nineteenth-century Europeans naturally thought that *all* swans were white until explorers discovered black swans in Australia. The same holds for scientific laws. No journey through the universe can verify that Newton's law of gravitation applies in all solar systems. Instead of seeking confirmation of the law, the task is to find counterexamples. As long as none are found, it will be assumed that Newton's law of gravitation holds. It is, as some readers may notice, a general inductive assumption. But it is not the inductive logic that Popper rejected. Kepler 'falsified' the Greek assumption of the circular orbit of planets. He replaced it by the hypothesis of elliptical orbits. Kepler's procedure

> ...(c)orresponds to the *method of elimination* which is applicable only if the theory is sufficiently easy to falsify – sufficiently *precise* to be capable of clashing with observational experience. (*Logik* 1935/^{5}1973: §39, p. 93; *Logic* 1959/1980: §39, p. 131; italics in original)

This example illustrates that falsification is a deductive relation between a universal theory and some test statements. (Popper called them 'basic statements'; the logical empiricists used the term 'protocol sentence'.) A basic statement (*Basissatz*) must be in contradiction with the universal statement. Throughout his career Popper affirmed that falsification was a deductive consequence of the clash between a universal statement and a singular basic statement. (*Logic* 1959/1980: §§1, 3, 6, 8, 18, 21; *Autobiography* 1974: §16) (Logically, this procedure is known as *modus tollens*.[8]) Deduction, he declared, is also 'an instrument of rational criticism'. (*Realism/Aim* 1983/1985: §27, p. 221) Universal theories are universal in scope, not limited in space and time. Hence, they cannot exhaustively be verified but they can be decisively falsified. But what is the difference between *falsification* and *falsifiability*?

3.2 Falsification and Falsifiability

Popper distinguished *genuine* from *illusory* falsifications, and *falsification* from *falsifiability*.[9] Falsifiability is a logical property of theoretical systems. They must be formulated in such a way that they can be refuted by empirical evidence. Their logical form must allow their potential falsification. Falsification is the act of falsifying a body of hypotheses. It reveals that a hypothesis or theory clashes with reality. This actual procedure requires rules, 'which will determine under what conditions

---

[8] This logical rule states that if in a conditional sentence: 'If p, then q', the consequent q does not hold, then the antecedent p must be negated. So we infer from non-q to non-p. If p stands for a theory and q stands for, say, a prediction, then the falsity of the prediction implies the falsity of the theory.

[9] *See Logic* 1980: §§3, 22; *Realism/Aim* 1985: xxii; *Alles Leben* 1996: 26; *All Life* 1999: 10; cf. Corvi 1997: Pt. II. In the Introduction to *Grundprobleme* (1979: XXXVI, 2009: XXXV; cf. C&R 1963: 228) Popper rejected the term 'falsificationism' because it conflated 'falsification' and 'falsifiabiliy'. He preferred the term 'fallibilism'.

a system is to be regarded as falsified'. These rules offer some scope: a single, irre-producible event should not be regarded as a falsifying instance. Scientists should also be aware that competing theories may be able to account for the occurrence of a predicted event. Theories may or may not be logically equivalent.[10] And theories, which make correct predictions—Greek geocentrism correctly predicted the orbital positions of planets—may not provide the correct explanation of the predicted event. Theories may also aim at explaining the existing evidence, rather than making new predictions. That is, science seeks both precise predictions and satisfactory explana-tions. Popper's insistence on the comparison of competing theories was an important feature of his methodology.

To see the logical property at work, contrast the statements 'It will rain' and 'It will rain *here tomorrow*'. Clearly, the first statement is too vague to be falsified. It will always be true at some place and some time. But the second statement is precise if *here* refers to a particular place and *tomorrow* to a particular time. This statement makes a precise prediction even if it is not a particularly exciting scientific one. Psychoanalysis was incapable of making such precise predictions even though it claimed to be a 'real' science. Eddington's prediction of 1935, however, that a solar eclipse would occur in parts of England in 64 years' time is a worthy scientific prediction.

> By a certain rule of inference, viz. the law of gravitation, we infer (..) the existence on Aug. 11, 1999, of a configuration of the sun, earth and moon which corresponds to a total eclipse of the sun. (…) The shadow of the moon on Cornwall in 1999 is already in the world of inferences.' (Eddington 1935: 92)

Although qualitative statements like 'All swans are white' are falsifiable, and indeed falsified, numerical precision increases testability. Scientific laws should therefore be numerically precise because this makes it easier to forbid and exclude many potential scenarios. Newton's inverse square law of gravitation—$F_g = G\frac{m_1 m_2}{r^2}$—excludes other possibilities, say $\frac{1}{r^3}$ or $\frac{1}{r^{2/3}}$. It commits the scientist to a precise quantitative relationship. Theories need to make precise quantitative predic-tions. This is one way of testing them. But more tests may be required if there are competing theories which fit the evidence. Medawar described this hypothetico-deductive conception of science as a dialogue between the possible and the actual. It is the invention of a possible world and the examination, through conjectures and refutations, whether the imagined world conforms with the real world. (Medawar 1984: 44–48)

The reader may now wonder about metaphysical statements. They can neither be verified nor falsified by empirical means. As Popper clarified in a letter to Lorenz (21/01/75), falsification only refers to 'the critical (not the dogmatic) part of knowl-edge' [321.04; my translation]. The falsification criterion only applies to empirical science. It does not apply to logic, mathematics or metaphysics. They follow their

---

[10] Popper dealt with such a situation in an article in *Nature* (1940). He discusses three interpretations of nebular red shifts: 'The three theories are logically equivalent, and therefore do not describe alternative *facts*, but the same facts in alternative *languages*.' ('Interpretation' 1940: 69–70; italics in original) (He would write further articles in *Nature* on the arrow of time in the 1950s and 1960s.)

own rules. Logic and mathematics obey logical proofs. Metaphysical views, such as determinism, are not devoid of meaning. They cannot be falsified but they can be criticized.

Scientists also make what Popper calls existential statements—claims such as 'there are white ravens' or more seriously there are 'black holes', 'gravitational waves' and 'dark energy and matter exist'. And science abounds with statistical statements. How do such statements fare under the falsifiability criterion? Although they are not strictly falsifiable, they must still be testable.

How can existential statements be tested? There is indirect evidence that various types of black holes exist, as well as dark energy and dark matter; just as there was indirect evidence of the fusion processes inside the interior of the sun. Assertions about black holes and dark matter are embedded in a theory, the General theory of relativity, which is testable. Popper stressed in *Logic* that the whole theory must be falsifiable. (1959/1980: Chap. III, §18) He counted existential statements as scientific if they belonged 'to a testable context'. Already in his *Grundprobleme* (1979: 317–318, 430–432; 2009: 478) and then in his *Logic* (§§18–22) and elsewhere ('Replies' 1974: 1035; *C&R* 1963: Chap. 10, §16; *Realism/Aim* 1983/1985: §22, pp. 187–188) he emphasized that falsification affects the 'whole system' and not only particular statements within it. Falsification is not a one-to-one refutation of every element in a theory. Theories are layered structures, which would make such a one-to-one refutation impossible.

> This structure allows us to distinguish between more risky or exposed parts of our theory, and other parts which we may—*comparatively speaking*—take for granted in testing an exposed hypothesis. (*Realism/Aim* 1983/1985: 186f, 178f; italics in original; cf. *Objective Knowledge* 1972: 359)

In *Logik,* Popper treated theories as universal statements ('All swans are white') when in fact they are complicated structures, which include empirical, mathematical, methodological and even metaphysical elements. They are in the words of Bronowski (1974: 628), who was sympathetic to Popper's fallibilism, 'topological networks'. Like Eddington, Popper described theories as 'nets'. They are human inventions to 'catch the world'.[11] But he also accepted, correctly in my view, that this kind of holism goes too far and that it is often possible to target particular hypotheses within a system. (*Poverty* 1957/²1960: Chap. IV, p. 132 [fn2]; *C&R* 1963: 238f) One example is Kepler's rejection of the assumption of circular orbits in Copernicus's heliocentric model.

Degrees of testability affect the whole system. Popper identified degrees of testability with degrees of empirical or informative content. ('Replies' 1974: 981–2; *Logic* 1959/1980: Ch. VI). He proposed an improvement of his demarcation criterion by specifying that 'a theory is scientific to the degree that it is testable'. Not only are there degrees of testability, there also exists Darwinian competition between theories.

---

[11] See K. Popper, 'On theories as nets', *New Scientist* (1982, 319–320). Popper repeatedly used this image of theories as nets, starting in *Grundprobleme* (1979: 487, 2009: 492). 'We try to examine the world exhaustively by our nets; but its mesh will always let some small fish escape: there will always be enough play for indeterminism.' (Popper, *Open Universe* 1982: 47)

What about statistical statements?[12] These are statements about radioactive decay or the therapeutic effectiveness of penicillin, even the average temperature in your region. Scientists, for instance, tell us that a certain element has a half-life of, say, n years. If the half-life of an element is n this indicates that out of a sample of X atoms of that element 50% will have decayed after n periods. The radioactivity of the sample loses half of its original intensity. Alternatively, it can be said that it is the time in which an atom has a 50% chance of disintegrating into another element. Different radioactive materials have different half-lives: 54.5 seconds for thorium but 4.51×10^9 years for uranium. (Weinberg 1993: 115–120) In Popper's view such statements are not verifiable because it can never be known exhaustively whether all or most elements have been observed. Are they falsifiable? Does the fact that half of a sample fails to decay after, say, 12 minutes (its half-life) 'falsify' the original statement? Medical practitioners may assure us that treatment with penicillin is effective in 95% of cases. If a patient does not respond to penicillin is the doctor's claim about its effectiveness thereby 'falsified'? Weather presenters tell us that it is 'unseasonably cool (or hot) for this time of the year'. Does this mean that the spell of cold or hot weather falsifies the statement about average temperature? In all these cases the answer is negative. Statistical claims, unlike strictly universal ones, are statements about certain percentages which will affect elements in a sample or population. It is not possible to predict which particular member of the sample or population—be it atoms or ravens—will satisfy the statistical prediction. Does this mean that falsification attempts must fail for such statements? If so, it would be a serious limitation of Popper's criterion because so many scientific theories are statistical in nature (evolutionary theory, quantum mechanics, and statistical mechanics). Popper agreed that, logically speaking, statistical hypotheses cannot be falsified. (*Logic* 1959/1980: §§44, 65, 68, 80) This is because they are not strictly universal. They do not state 'all Xs are y' but 'a given percentage of Xs are y'. Unlike universal statements they allow for exceptions. A colony of black swans does falsify the claim that *all* swans are white. But the discovery of a colony of black swans would not per se falsify the statement that *most* swans are white. It depends on the size of the respective colonies. So even if ornithologists (bird biologists) observed a large number of black swans this would not, logically speaking, constitute a counterexample to the ornithological thesis that 'most swans are white'. But how much credibility would this purely logical point enjoy amongst scientists? Very little! Popper therefore concluded that statistical statements are only *practically* falsifiable. Consider the radioactive decay of elements. Let the claim be that the half-life of an element is, say, 24 hours. Laboratory tests, however, consistently show that it is much shorter or much longer. Scientists would then adopt a conventional rule, which would allow them to regard the original estimate as practically falsified. The rule does not forbid the occurrence of deviations or atypical events. But it does forbid the 'predictable and reproducible occurrence of systematic deviations.' (*Logic* 1959/1980: §68; Gillies 2000: 145–150)

---

[12] Popper's concern with probability in *Logik* later led to his well-known propensity interpretation of probability.

3.3   On the Conflation of 'Falsifiability', 'Refutability', and 'Testability'

Throughout his writings, Popper makes equivalent use of such notions as 'falsifiability', 'refutability' and 'testability'. A typical example occurs in *Conjectures and Refutations* (1963: 36–37; italics in original), where he states that

> … every genuine *test* of a theory is an attempt to falsify it, or to refute it. Testability is falsifiability; but there are degrees of testability: some theories are more testable, more exposed to refutation, than others; they take, as it were, greater risks.

But when it comes to statistical statements, he relaxes his criterion of strict falsifiability to practical falsifiability. If this is the case, it would be better, as some commentators have argued (Kraft 1974; Maxwell 1974; Keuth 2005) to speak of *testability* of scientific theories, rather than falsifiability. Such a criterion of practical falsifiability or *testability* is in closer accord with empirical science than strict falsifiability. Before turning to the question of the compatibility of Popper's methodology with empirical science, let us consider other reasons why the notion of testability is better suited than falsifiability.[13]

One reason is that Popper reacted very strongly against accusations that he was a 'naïve falsificationist'. One such suggestion came from the American physicist and historian of science Thomas S. Kuhn, the author of the famous book *The Structure of Scientific Revolutions* (1962). Although Kuhn (1974: 14) granted that Popper was not a 'naïve falsificationist' he suggested that he could legitimately 'be treated as one'. Lakatos (1974: 93) claimed that 'Popper's logic of scientific discovery' conflates 'naïve' and 'sophisticated falsificationism'. Naïve falsificationism means that a theory, which has been refuted by evidence, should be abandoned. Scientists should go in search of a better theory. It is true that Popper sometimes sounds like a naïve falsificationist, especially in his early work. For instance, even retrospectively, he describes falsification as *modus tollens.* (*Autobiography* 1974: 62) But he never really was one. In a letter to Feigl (05/06/61) he insisted that he had never spoken of 'definite falsification' but always of the '<u>non</u> finality of falsification' [294.06; underlined in original]. When this accusation was raised during a panel discussion in the 1980s, he rejected it out of hand. (*Zukunft* [4] 1990: 56–58; *Realism/Aim* 1983/1985: Introduction 1982: xxxiii; *Logic*: 1959/1980: §6)

Why was Popper not a naïve falsificationist? In *Grundprobleme* (1979: 163; 2009: 143) Popper anticipated a theme, which became dominant in his later work. It is a comparison with Darwinism: species fight for survival, competing theories fight for acceptance. Selection is the methodological mechanism of scientific progress. His *Logic* (§30; cf. §82 [fn 2]; cf. *Poverty* 1957/[2]1960: §27) contained a further hint, which without hindsight could easily be missed. Popper invoked the *selection* of

---

[13] This is not just an issue of terminology. The German sociologist Ulrich Beck uses Popper's criterion of 'practical fallibilism' as an element in his theory of the 'risk society', because it undermines the traditional image of science, which Popper himself rejected. (Beck 1992: Pt. III, Chap. 7)

competing theories through tests.[14] Theories, like biological species, compete. The emphasis is on the 'natural selection of scientific hypotheses and the struggle for the existence of theories'. They aim to become the dominant, accepted 'paradigm'. Why would such a Darwinian influence ban the spectre of naïve falsificationism? First, it is a fact in the history of science, that very often more than one explanation is available. One theory may become dominant, like Newtonian mechanics or Darwinian evolution, but it remains in competition with alternative or even rival theories.[15] Popper himself wrote at length about the struggle between various theories of materialism. (Popper and Eccles 1977: Part I) But there are more noteworthy examples. Nicolaus Copernicus consistently compared his own theory of heliocentrism (1543) with the then dominant theory of geocentrism. He concluded that his heliocentric model fitted the evidence, both mathematically and empirically, better than the Ptolemaic model. In 1651 the Italian astronomer, Giovanni Battista Riccioli, published his influential textbook *Almagestum Novum* ('The New Almagest'). It presented five alternative or rival theories of the planetary system, including the Copernican one. Charles Darwin consistently defended his own theory of evolution against both the rival of creationism and the alternative of Lamarckian evolution. So, when a theory faces recalcitrant evidence, there will be competing theories, which strive to account for it. It will then be necessary to find further evidence, which can weed out these alternatives or rivals until the most plausible explanation can be found. Popper never claimed that a theory should be quickly abandoned when facing falsifying evidence. A theory may be falsified but it is only rejected when a better alternative becomes available. In *Logic* (§11), he declared that a corroborated hypothesis should not be abandoned without good reason. (For the moment, read 'corroboration' as a replacement for Carnap's 'confirmation'; the difference will be discussed later.) Some degree of 'dogmatism', as he later admitted more openly, was necessary to give a theory a chance to parry contrary evidence. 'There can be no critical phase without a preceding dogmatic phase'. There are only better or worse theories and falsification is never conclusive.[16]

---

[14] On the question of proliferation of hypotheses, David Miller told me that 'he (Popper) had learnt from his geologist colleague Bob Allan in NZ about Chamberlin's paper 'The Method of Multiple Working Hypotheses', which was published in the *Journal of Geology* ([5]1897: 837–48, and reprinted in *Science* in 1965 http://science.sciencemag.org/content/148/3671/754). Jeremy Shearmur procured him a copy [349.13].

[15] I understand the difference between alternative and rival theories as that between *alternative* versions of the same theory, which agree on first principles, and *conflicting* theories, which disagree on first principles.

[16] Popper frequently stressed the importance of a dogmatic phase, not only in his publications— *Autobiography* 1974: §§10, 16; 'Replies' 1974: 984; *Myth* 1994: 16; *Alles Leben* 1996: 121; *All Life* 1999: 41; *Realism/Aim* 1983/1985: Introduction 1982: xxii—but also in his correspondence. In a letter to the American physicist and philosopher Abner Shimony (01/02/70), whom he met at Brandeis, he emphasized that, against the slogan of verification, he had to stress the 'virtues of testing'. He added that "dogmatic thinking" and the defence of a theory against criticism are needed, if we wish to come to a sound appreciation of the value of a theory: if we give in too easily, we shall never find out what is the strength of the theory, and what deserves preservation'. Not happy with

Consider geocentrism. The brilliant Greek mathematician and astronomer Ptolemy proposed his influential earth-centred theory of the planetary system, according to which planets move around the central earth in *circular* orbits. The earth sits 'motionless' at the centre of the 'universe'. All other planets, as well as the sun, orbit around it. The idea of circular orbits held a central place in Greek astronomy. At the same time the Greeks knew from well-established observations that planets do not follow circular orbits. Planets are sometimes nearer, at other times further away from the earth. Yet the Greeks did not use these observations to reject their central dogma: the circular motion of the planets. Still, geocentrism is regarded as one of the greatest achievements in the history of astronomy. It ruled supreme (with very few exceptions) until Nicholas Copernicus challenged it (1543). But Copernicus retained the traditional dogma of circular motion. The Copernican heliocentric view took over 150 years to establish itself. One would hardly call it conjectural knowledge that the earth is one of nine planets (including Pluto), all of which orbit the sun in ellipses of various degrees of elongation. Popper displayed a somewhat ambiguous attitude: he praised the boldness of Copernicus (and his Greek predecessor Aristarchus of Samos) for suggesting that the earth was not the centre of the universe. Insofar as the Copernican theory made a number of 'minor predictions', he considered it 'scientific'. But as neither Aristarchus nor Copernicus discovered or predicted new astronomical facts, he branded their theories as 'unscientific or metaphysical'. ('Replies' 1974: 978; cf. Weinert 2010)

Modern cosmology is in a similar situation today. String theory attempts to combine quantum mechanics and relativity to create a quantum theory of gravity. It postulates that at the elementary level, particles—hadrons which consist of quarks and anti-quarks—can be broken down into strings. But strings exist at the unobservable Planck scale (10^{-35} m). Opponents therefore reject String theory as unfalsifiable speculation. It is not just the unobservable minute scale at which strings are supposed to exist. The theory also requires many more than three spatial dimensions to be mathematically consistent.

What is needed is some dogmatism and a lot of competition. In response to Kuhn's challenge (1962), Popper returned to the evolutionary theme regarding the growth of scientific knowledge, the roots of which lie in his emphasis on the natural selection of theories. (*Logik* 1935/⁵1973: §30, p. 73; *Logic* 1959/1980: §30, p. 108) It encourages the elimination of theories by the method of trial and error, as he explained in a paper to a philosophy seminar at Canterbury University College in Christchurch in 1937. (It was published under the title 'What is Dialectic', in *C&R* 1963: Chap. 15.) His interest in evolutionary thinking appeared again in his Preface to the first English edition of *Logik* (1959/1980: 15; italics in original). There he declared that the

> …(c)entral problem of philosophy has always been and still is the problem of the growth of knowledge. *And the growth of knowledge can be studied best by studying the growth of scientific knowledge.*

Popper's version of fallibilism, Shimony hoped to persuade him of the power of scientific inference [350.07].

In his later work, his evolutionary view of the growth of scientific knowledge was captured in a *tetradic scheme*, according to which theories try to solve problems by error elimination (*Objective Knowledge* 1972: 243):

$$P_1 \rightarrow TS \rightarrow EE \rightarrow P_2$$

(Old) Problem$_1$ $\rightarrow$ Tentative Solution(s) $\rightarrow$ Error Elimination $\rightarrow$ (new) Problem$_2$.

According to this scheme, science evolves from the perception of a problem to the emergence of a new problem, via tentative solutions and error eliminations. Nevertheless, it is a simplified scheme, which is made more realistic by taking the force of selection amongst alternative and rival theories more seriously. This multiplies the tentative solutions. Already this simplified form shows that by the logic of Popper's own system the notion of testability captures the spirit of science more accurately than strict falsifiability.

3.4 Against Ad Hoc Manoeuvres

According to Popper, disciplines like Freud's psychoanalysis, Adler's psychology and Marx's economics display a tendency to save their theories against the threat of falsification. A popular device is the introduction of ad hoc measures to protect a theory against refutation. Popper employed a fictitious example to illustrate such ad hoc defences. (*Objective Knowledge* 1972: 192)

> Why is the sea rough today?
> Because (God) Neptune is angry.
> How do you know that Neptune is angry?
> Do you not see that the sea is rough?

Here, in Popper's words, the only evidence for the *explanans* ('Neptune is angry') is the *explanandum* ('the sea is rough'). In other words, there is no independent evidence of Neptune's anger. In order to avoid circularity, such independent evidence is clearly important. Freud, however, failed to provide it for his central doctrine of the Unconscious. The only evidence he offered for the explanans (the Unconscious) was the explanandum itself (Freudian slips, weird dreams and neurotic behaviour). By analogy with Popper's example:

> Why did the patient commit a slip of the tongue?
> It is the work of the Unconscious.
> How do you know that the Unconscious is responsible?
> Did you not hear the Freudian slip?

Popper was right that the central hypotheses of a theory must not be protected against possible refutation by ad hoc manoeuvres. But even if ad hoc modifications can be avoided is falsification too blunt an instrument, as one scientist put it, to be useful in science?

3.5 What Scientists Think of Popper's Falsifiability

As Popper always emphasized the applicability of his philosophy, the issue of what scientists think of falsifiability, should be addressed. There are those who agree with Popper, for instance, the neuroscientist John Eccles, whom Popper met in New Zealand. He later co-authored their book *The Self and Its Brain* (1977). Popper converted Eccles, the Nobel Prize winner in Physiology or Medicine in 1963, to the deductivist way of thinking.

> Until 1945 I held the conventional ideas about scientific research. First, hypotheses grow out of the careful and methodical collection of experimental data, according to the inductive view deriving from Bacon and Mill.

In 1974, when he wrote these lines, he still believed that many scientists followed this inductive procedure. Secondly, he believed that 'the excellence of a scientist is judged by the reliability of his developed hypotheses, and by how much they stand firm as secure foundations for further advance'. And then came the most important point:

> It is a sign of failure and in the highest degree regrettable if a scientist espouses an hypothesis which is falsified by new data and has to be scrapped. (Eccles 1974: 347)

Peter Medawar (1915–1987), Immunologist. *Source* Wikimedia Commons

Scientists like Eccles adopted Popper's rejection of the inductive method in science. Medawar, who won the Nobel Prize in Physiology or Medicine in 1960, shared this view. 'Induction is a myth' he states in his book *Pluto's Republic* (1984:

117; cf. Levinson 1974; Deutsch 1997: Chap. 7). But, as he wrote to Popper, 'inductive fallacy dies hard' (22/12/58) [325.26]. Like Magee, Medawar did much to spread the message and promote Popper's ideas amongst scientists. He even promised to write a piece on Popper's philosophy in *Vogue* magazine (letter to Popper September 7, 1973) [325.27]. Popper's supporters thought that his emphasis on falsifiability kept scientific dogmatism at bay.

How should we think about speculative theories? The physicist Lee Smolin, of the Perimeter Institute (Canada), dismisses the so-called Anthropic Principle[17]—defended by String theorists—as unscientific because

> …it fails to have a necessary property to be considered a scientific hypothesis. This is that it be falsifiable. According to [the philosopher] Popper a theory is falsifiable if one can derive from it unambiguous predictions for doable experiments such that, were contrary results seen, at least one premise of the theory would have been proven not to apply to nature. (Quoted in Susskind 2006: 192)

Leonard Susskind of Stanford University, one of the originators of String theory, regards such demands for falsifiability as 'overzealous Popperism' (2006: 195). He quotes a number of examples from the history of science—from Darwinism via atomism to quark theory—whose premises were dismissed as unfalsifiable only to be accepted later. For Susskind falsification—I think he means 'strict' falsification—is a 'red herring'. Defenders of 'strict' falsifiability are in his eyes 'Popperazi' (2006: 192). The cosmologist Sean Carroll also rejects falsifiability:

> The falsifiability criterion gestures towards something true and important about science but it is a blunt instrument in a situation [he is referring to the multiverse] that calls for subtlety and precision. It is better to emphasize two more central features of good scientific theories: they are *definite*, and they are *empirical*. (Carroll 2014: 2; italics in original)

Leonard Susskind (*1940), cosmologist. *Source* Wikimedia Commons

---

[17] Some of the leading proponents of string theory also embrace the Anthropic Principle. (Susskind 2006: 197) It does not just claim that the world is the way it is *because* we are here. No, the Anthropic Principle serves to explain the fine-tuning of the constants of nature, without which (intelligent) life would be impossible.

Carroll agrees with Popper that science should make 'definite' and unambiguous statements about the external world. But by 'empirical' he does not mean that science should simply make 'falsifiable predictions', because 'the interplay between theory and experiment is not so cut and dry'. Rather, a 'scientific theory should be judged by its ability to account for the data', which can be done directly or indirectly. The Nobel Prize winning physicist Steven Weinberg (Nobel Prize 1979) hailed Popper as 'the dean of modern philosophers of science'. (Weinberg 1993: 165) Yet he found no single instance of falsification 'in the past one hundred years'. (Weinberg 1993: 102) Popper disagreed. In the Introduction to *Realism and the Aim of Science*, written in 1982, he listed twenty examples of 'refutation', going back to the Greeks. His examples include Rutherford's refutation of the Thomson's plum-pudding model of the atom[18]; the failure of Newton's theory of long-range gravitation to explain and predict the full effect of the perihelion advance of Mercury, which led to Einstein's field theory of gravitation; and the failure of classical physics to account for atomic spectra resulting in Planck's quantum hypothesis.

Eccles and Medawar were not the only scientists who agreed with Popper. The famous evolutionary biologist, Stephen Jay Gould (1984: 256), appealed to Popper's falsifiability criterion to dismiss creationism as unscientific. The physicist David Deutsch, of Oxford University, agrees with Popper that science is a problem-solving process. (Deutsch 1997: Chap. 3, p. 62) He professed in a letter (04/08/1977) that he was a 'staunch Popperist', although he did not agree with his 'realistic interpretation of quantum mechanics' [288.03]. The unorthodox American physicist David Bohm also concurred with Popper on falsifiability. (In 1961 Bohm became professor at *Birkbeck College*, London, not far from the *LSE*.) Chapter XXV of his book *The Special Theory of Relativity* (1966) is devoted to the topic of 'The Falsification of Theories'. He writes:

> It is clear that the falsification of old theories has actually had a key role in the development of physics (and indeed of the whole of science). A little reflection on this problem shows, however, that such a process is a necessary part of the development of a science, and that in order to permit this development to take place properly, it is indeed essential that scientific theories be *falsifiable*. (Bohm 1965: 123–124; italics in original)

This attitude implies further, again in agreement with Popper, that absolute certainties do not exist in science. The well-known British-Austrian cosmologist Hermann Bondi opens his book *Assumption and Myth in Physical Theory* (1967) with a chapter on Popper's 'method of disproof'. Bondi, who was born in Vienna, found himself in full agreement with Popper.[19] On occasion of Popper's 90th birthday Bondi published a short piece in *Nature* (1992), which celebrated 'the philosopher for science'. He

---

[18] Joseph J. Thomson proposed the 'plum-pudding' model in 1904, after his discovery of the electron (1897). The negatively charged electrons were embedded in a positively charged volume, but there was no nucleus. It was replaced by Rutherford's nucleus model. For more on these models see my book *The Scientist as Philosopher* (2004) and my articles 'The Structure of Atom Models' (2000) and 'The Role of Probability Arguments in the History of Science' (2010).

[19] Bondi is famous for his contribution to cosmology. He rejected the Big Bang theory and proposed, in cooperation with Fred Hoyle and Thomas Gold, the alternative steady-state model. Fred Hoyle's biographer Simon Mitton, of Cambridge University, told me in a private email (06/03/2020) that

praised Popper's emphasis on problem-solving 'as the central activity of humans', which 'describes so particularly appositely what is done in science'. He confessed that Popper's fallibilism—'the idea of theories having to be vulnerable to empirical disproof'—profoundly influenced him and many others.

The Nobel Prize winning economist Milton Friedman (1976) also echoed Popper's fallibilism, when he argued that a 'hypothesis is rejected if its predictions are contradicted ...; great confidence is attached to it if it has survived many opportunities for contradiction. Factual evidence can never "prove" a hypothesis...'. (Friedman 1953: 9; cf. Parvin 2013: 125) (Note, however, that Popper disapproved of Friedman's economic instrumentalism, *C&R* 1963: 102, 245; *Myth* 1994: 182 [fn 6]).

Sometimes, scientists are 'Popperians' without being aware of it. Jim Al-Khalili, the British physicist and television presenter, makes a number of Popperian statements in his book *Black Holes, Wormholes, and Time Machines* ([2]2012)—for instance that 'we can never prove a theory only disprove it'. But he told me in a private email (13/06/2018) that such statements are not 'intentionally Popperian'.

> They are just how I (and, I suspect, many scientists[20]) think is a good definition of the scientific method. So, if they are Popperian ideas, then yes I would say they are widely accepted and acknowledged in science, particularly in fundamental physics.

This is certainly how Paul Nurse, the Nobel-winning geneticist (2001), thinks of the scientific method. In a lecture at the *LSE* (podcast 2016), he was explicit about his debt to Popper's ideas. Popper helped him, as he helped John Eccles, to shift his view of science from an inductive, data-gathering approach to a deductive, trial-and-error approach. Popper's methodology provided scientists with a compass to help them deal with their experimental data. But Nurse also pointed out some limits of Popper's methodology. Scientific research must leave room for data gathering *with* the aim of formulating a hypothesis. Accidental discovery cannot be ruled out. In some situations, he recalled from his own experience, it is possible to proceed without a clear hypothesis in mind, although this requires relevant research. Its relevance may be judged by previous research. As Popper's approach is normative, he failed to consider the messiness of experiments and scientists.[21]

Although Carroll and Susskind express reservations about (strict) falsifiability, they accept of course the need for theories to be tested. But such tests can be either direct or indirect. Foucault's pendulum (1851) provided direct evidence that the earth turns on its own axis. Whilst there is no such direct evidence of the nuclear processes

---

Hoyle never mentioned Popper. Popper dismissed the Big Bang theory as 'unimportant' (*Offene Gesellschaft* 1986: 48–50), even as 'metaphysical'. (*Zukunft* [4]1990: 69–70)

[20] For instance the great American physicist Richard Feynman who held that science is not certain, that it starts with 'guesses' whose consequences must be compared to experience.

[21] In our conversation at the *LSE* John Worrall sounded a note of caution with reference to Peter Medawar and Paul Nurse: 'well, quite honestly, I don't know whether you really need to read Popper to know pretty soon when you are doing your scientific work that you are not inductively generalizing data, that you do make hypotheses, that you do need to check that these hypotheses are true or not'. But he agreed that 'far and away more than any other philosopher he does seem to have been generally influential. And generally regarded as a significant figure, more outside the field than within the field, I think'.

in the interior of the sun, there is indirect evidence. For instance, in 1859, Gustav Kirchhoff and Robert Bunsen developed spectral analysis. (Weinert 2004: 2009–2010) By analysing the spectrum of electromagnetic radiation of sunlight, they were able to establish that the sun contains sodium. Spectral analysis also revealed the atomic constitution of the sun. Knowing that the sun consists mainly of hydrogen and helium scientists can reconstruct the process of nuclear fusion by which the sun produces the energy to sustain life on earth. Today, spectroscopy is also employed to study the chemical composition of cosmic particles, and which processes led to them since the Big Bang. Popper himself stated that indirect measurements 'are practically the only measurements' that occur in physics. (*Logik* 1935/^{5}1973: §75, p. 176 [fn4]/*Logic* 1959/1980: 226 [fn4])

Another form of indirect evidence consists of mathematical deductions. If Newton's laws are correct, then it is possible to work out the mass of the earth. The earth cannot be put on a scale—nor can a galaxy for that matter. However, by equating Newton's second law ($F = ma$) with his gravitational law, it is possible to derive, in a few steps, the mass of the earth, m_E.[22] Both Weinberg and Susskind mention the mathematical consistency of a theory as indirect evidence. However, mathematical consistency can only be a necessary not a sufficient condition. Ptolemaic astronomy was a mathematically consistent theory, but it was empirically false. String theory, as I understand it, seems to resemble Ptolemaic astronomy in this mathematical respect. Susskind (2006: 293) reminds his readers that mathematical existence is not to be confused with physical existence.

Even sceptical scientists, such as Carroll and Susskind, will accept 'falsifiability' if it means testability by direct and indirect means. They have the grand theories of cosmology in mind. The physics Nobel Prize (2020) was awarded for the demonstration that a black hole exists at the centre of our Milky Way. This supermassive black hole, which occupies a region the size of the solar system, remains invisible. But its existence is inferred from the observation of the extremely rapid rotation of stars near the centre of the galaxy. What about scientific work at a more mundane level? One scientist was intrigued by the question of why some pebbles on beaches display a nearly perfect elliptical shape. The water flow carries them back and forth on a sandy beach, where they are subject to friction and rotation. Only stones which rotate around the axis of greatest moment of inertia will form into an ellipsoid. Using only Newtonian mechanics he set about working out the mathematics, which determines their smooth, rounded shape.[23] He collected more than 1250 pebbles, on which he carried out his measurements. He estimated that material with hardness similar to basalt will form into an ellipsoid within 500–1000 years. Was he in the business of falsifying his theory? Did he submit it to severe tests? No, because the facts were already well known. It was a question of fitting the mathematics to the known facts.

---

[22] Equate Newton's second law of motion and his law of gravitation: $mg = G\frac{mM_E}{r^2}$ and solve for M_E. Here g is the acceleration near the surface of the earth, r is the radius between the centres of the two bodies and G is the gravitational constant.

[23] Winzer (2019); cf. Kneale's example of Anderson's discovery of the positron. Kneale (1974: 206–208). Settle (1974: 701–702) discusses some further examples of 'non-Popperian' progress in science.

Popper encouraged the view that 'scientific success consists of unsuccessful attempts at refutation'. ('Replies' 1974: 996) He was fond of claiming that all our knowledge is conjectural, based on trial and error. It followed from his insistence on testability. But if testability comes in degrees, as he stated (*Logic* 1959/1980: §§31–40), then so does conjectural knowledge. (Note that Chapter VI of *Logic* is entitled 'Degrees of Testability', a translation of 'Grade der Prüfbarkeit'). Some areas of knowledge are less conjectural than others. It is usually the cutting edge of research which is more conjectural than established knowledge. Although the Copernican heliocentric theory was at one point highly conjectural, it is established knowledge now that the planets perform elliptical orbits around the sun. By contrast String theory, with its many-dimensional universe, is far more conjectural. It is much more likely to be overthrown than heliocentrism.

Given these considerations it is probably wise to reconsider Popper's conflation of falsifiability = refutability = testability. Falsifiability and refutability are equivalent terms. But testability, understood as direct or indirect testing and the weighing of competing theories, better fits scientific practice—both by the logic of Popper's methodology and the practice of science. (Box 1) The Popperian claim that fallibism is the norm has made his readers overlook an important section in *Logic*, entitled 'Deductive Testing of Theories' (Chapter I, §3). In this paragraph Popper spells out that falsifiability is only one part of a wider notion of testability.

> We may if we like distinguish four different lines along which the testing of a theory could be carried out. *First* there is the logical comparison of the conclusions among themselves, by which the internal consistency of the system is tested. *Secondly*, there is the investigation of the logical form of the theory, with the object of determining whether it has the character of an empirical or scientific theory, or whether it is, for example, tautological. *Thirdly*, there is the comparison with other theories, chiefly with the aim of determining whether the theory would constitute a scientific advance should it survive our various tests. And *finally*, there is the testing of the theory by way of empirical applications of the conclusions which can be derived from it. (*Logik* 1935/^{5}1973: 7–8; *Logic* 1959/1980: 32–33; italics added; *Objective Knowledge* 1972: Chap. III, §8; cf. Settle 1974: 721; Simkin 1993, Pt. I, §7)

And if there are strong methodological similarities between the natural and the social sciences, as Popper claimed, testability is once again a better term. Social science theories can make predictions, so they can be tested. But for reasons to be explored below they are hard to submit to direct falsifications.

Box 1: Popper's Examples of Refutations in the History of Science

In the light of this result, it is worth reconsidering Popper's 'List of Examples Chosen Almost at Random'. (*Realism/Aim* 1988: xxvi–xxx) Popper starts his discussion by equating 'falsifiability' with 'empirical refutability'. But on his list only a few cases are clear examples of refutation by empirical means: Lavoisier refuted the phlogiston theory of combustion; Young's two-slit experiment disproved Newton's corpuscular theory of light and Rutherford's large-angle scattering experiment eliminated Thomson's 'plum-pudding' model of the atom. [He could have stated more clearly than he did in *Quantum Theory/Schism* (1982/1992: 24–25) that Bell's inequalities falsify a whole class of local hidden variable theories. He could also have added the celebrated Stern–Gerlach experiment of 1922, which was a clear refutation of the classical view of the behaviour of atoms in magnetic fields, and established the quantization of angular momentum of atoms.] Interestingly, Popper proposes that the Davisson–Germer experiment (1927), which was the first measurement of the wavelengths of electrons, refuted Schrödinger's interpretation of quantum mechanics. He does not spell out which one, for Schrödinger changed his mind several times. It suggests, however, that some interpretational issues are not just criticizable but falsifiable. Galileo, however, used a thought experiment to show the inconsistency in Aristotle's explanation of falling objects. Toricelli discovered vacuum in a tube of mercury, which could not have refuted, by Popper's own standards, the metaphysical doctrine that 'nature abhors the vacuum'. It makes no quantifiable prediction. 'Nature' is not a quantifiable term; 'abhorrence' is not an empirical property. Boyle concluded that such a metaphysical doctrine should have no place in empirical science where quantifiable explanations should be employed. Kepler's falsification of circular planetary orbits was not a case of a direct test of the Copernican theory. It was at best a test of a core part of the theory. Kepler saw it as a refinement of the heliocentric model, which better fitted Tycho Brahe's empirical data (for instance, the observation of eccentric orbits of comets). Other examples on Popper's list are also cases of modification rather than direct refutation. Thomson's discovery of the electron (1897) was a refinement of the ancient view that atoms are the ultimate constituents of nature. Anderson accidental discovery of the positron (1932) was not a deliberate attempt to disprove the 'two particle model' (electrons and protons) of the atom.

Going beyond Popper's examples, the famous Michelson–Morley experiment (1897) was an attempt to detect the motion of the earth with respect to a hypothetical material ether, not an attempt to overthrow the ether hypothesis. (Bernays 1974: 599) Precise tests of Einstein's equivalence principle (between inertial and gravitational mass) are not attempts to overthrow but to consolidate this fundamental principle. Mendeleev's discovery of the periodic table (1869) was not an attempted refutation nor is the search for exoplanets or water on Mars.

Popper concludes that the 'theory of refutation followed by revolution and yet conservative reconstruction' throws much light on the history of science. It confirms, I think, my reinterpretation of the principle of falsifiability as a principle of testability. Falsifiability is a criterion of the empirical character of scientific theories but there are more ways of testing a theory than confronting it with its deductive consequences. In a reply to Bartley's 1965 attack on his demarcation criterion, Popper himself granted that 'empirical refutation is just a part of the general critical work of improving theories' [271.22]. And he proposed logical and material criteria for the potential and actual progress of one theory over another (see below Chap. 6.4).

3.6 Deductivism, Inductivism and Testability

Popper rejected the verifiability criterion of the *Vienna Circle* as a criterion of meaning *and* of demarcation. The latter criterion was based on the predominant view, as highlighted by John Eccles and Paul Nurse, that science proceeds by induction, a method Popper rejected. Popper's anti-inductivist view, which set him on a collision course with the *Vienna Circle*, arose from the asymmetry of verification and falsification. Popper fully accepted Hume's famous criticism of induction. Induction is the tendency (a) to generalize from a few cases to all cases: from 'some swans are white' to 'all swans are white'[24]; (b) to infer from past to future cases: 'the sun has always risen in the East, so it will rise in the East tomorrow'; (c) to infer from past cause–effect relationships to future cause–effect relationships: from 'this light switch has worked so far, to 'it will work today, tomorrow…'. This procedure is dubbed 'induction by enumeration' because it adds up the evidence step by step.

As Hume, the Scottish Enlightenment philosopher, pointed out there is neither an empirical nor a logical proof to justify an inference from the observation that 'some p' to the universal statement that 'all p'. His first argument was of an *empirical* kind. There is no empirical reason to infer from the particular to the general or from past to future cases. It cannot justifiably be reasoned that 'it happened in the past' so 'it will happen in future'. The only evidence we have for this claim is our experience of past cases. By inferring from the past or the particular to the future or the general, we are presupposing the validity of induction for which the only evidence is what we have observed in the past.

His second argument was of a *logical* kind. It concerns the difference between a *deductive* and an *inductive* argument. In a deductive argument the truth of the premises carries over to the truth of the conclusion. If it is true that 'all planets orbit the sun' and that 'Mars is a planet' then it follows logically that 'Mars orbits the sun'. But in inductive arguments the truth of the premises does not guarantee the truth of the conclusion. The premises can be true, but the conclusion is still false. Consider the premises: 'The sun rose every day from the distant past to the recent past', 'the sun rose every day of my lifetime', 'the sun rose yesterday'. All these premises are true but it does not follow from the truth of the premises that the conclusion: 'the sun will rise tomorrow' is true. This feature captures the logical nature of inductive arguments. The truth of the premises does not confer truth on the conclusion.

---

[24] Note that national or racial prejudices are based on inductive steps: from our experience with some people of a nation or a race to all people of that nation or race.

In *Logik*, Popper concluded from this state of affairs that it was impossible to establish the truth of a universal scientific theory on the basis of a limited number of observations. Universal theories in science cannot be established by inductive reasoning. Popper stood this situation on its head. The truth of a universal theory cannot be verified by observation or experiment. But it can at least be shown that a theory is false. If a universal theory is cast in such a form that it may be at odds with some observation statements, from which they are derived, then there is reason to believe that the theory is false. This is the asymmetry of verification and falsification. A theory in science cannot be verified but it can be falsified. It is falsified when some of its deductive consequences contradict the empirical evidence. (Popper 1959, 1963, 1973) Kepler's first law of elliptical planetary orbits was a discovery that falsified the presupposition of circular orbits in Ptolemy's geocentrism and Copernicus's heliocentrism. The discovery of elliptical planetary orbits led Kepler to a rejection of geocentrism *and* a modification of Copernican heliocentrism. The elliptical orbits take the sun, not the earth, as a focal point. In reality Kepler's problem situation was more complex. He formulated his planetary laws on the basis of Tycho Brahe's observations. Tycho Brahe, a Danish astronomer, developed his own planetary model, which he derived from his observations. Brahe kept the earth as the centre of the 'universe' with the moon and the sun orbiting the earth. But the planets themselves orbited the sun. Kepler's modification of the heliocentric system also implied the rejection of Brahe's compromise model. There was thus competition between rival astronomical models. Kepler's discoveries did not immediately lead to the victory of the Copernican system. Its final endorsement only occurred after Newton's *Principia* (1687).

If inductive reasoning in everyday life stands on shaky ground, how much worse will it be in scientific practice? Popper transferred Hume's scepticism regarding ordinary to scientific inferences. In analogy to Hume he asked: Is there ever any reason to infer from the truth of particular observation statements to the truth of scientific theories? His answer was 'no'. The reason for his negative answer was once again the asymmetry between falsification and verification. Newtonian mechanics applies to *all* macro-objects—objects of a particular size—which move at ordinary velocities. Before the age of space travel non-existent rockets were subject to Newton's laws. These laws do not just cover all seen objects, say planets, they also apply to all unseen objects. They apply to hypothetical planets, too. Newton's law of gravitation is universal.[25]

Despite this inductive scepticism it is hard to do without inductive assumptions, both in everyday life and in science. In ordinary life, for instance, people assume that the trains will run and the shops will be open. In science, it is assumed that planets in far-flung reaches of the universe obey Newton's law of gravitation. Popper in fact accepted that inferences from particular to numerically general statements are

---

[25] Note that Newton's theory does not require that all planets rotate from west to east. In our solar system both Venus and Pluto spin from east to west. So, the east-bound spin of most planets in the solar system could not be a universal, all-inclusive law.

legitimate. They are just finite conjunctions of singular statements. (*Logic* 1959/1980: Chap. III, §13) He objected to induction as a method of scientific investigation.

Having accepted Hume's analysis, Popper turned Hume's question on its head. Can we infer from the truth of some observation statements—test statements—the 'falsity' of a universal theory? His answer, as we have seen, was decidedly 'yes'. To refute a universal theory it is sufficient, logically, to observe some negative instances (some black swans or some planets which disobey Newton's laws.) Hence Popper arrived at his idea of falsification—in contrast to inductive verification. Given the weight of the inductive tradition this was a revolutionary step in the philosophy of science of the 1930s. Popper put the emphasis on the primacy of theory, not data, and the primacy of problems. It is the Kantian theme, to which he was indebted. (The overwhelming influence of Kant's philosophy emerged in *Grundprobleme* and it stayed with Popper till the end.) Popper turned this primacy-of-theory-view into a methodology of scientific research. Universal scientific theories are not inductively built up from observation statements. Popper had a powerful ally in Einstein. Einstein was right to stress that the complicated equations of the General theory of relativity cannot be obtained by summing over observation statements.

> I have learned something else from the theory of gravitation: No even so inclusive collection of empirical facts can ever lead to the setting up of such complicated equations. A theory can be tested by experience, but there is no way from experience to the setting up of a theory. (Einstein 1949: 89)

The same is true of the Special theory of relativity. Its equations were the result of pure thinking, against a background of acquired physical knowledge. They were then subjected to tests.

Anti-inductivism was one element of Popper's revolution. It figured prominently in the pages of his *Grundprobleme* and his *Logik*. There were two further elements: his rigorous objectivism and his unfailing anti-psychologism. (*Open Society* II, 1945/⁵1966: Chap. 14; Musgrave 1974) His objectivism expressed itself in the manner he envisaged the rational assessment and testing of theories as well as his insistence on the objectivity of probability. His anti-psychologism featured prominently in the distinction between 'subjective experience' and 'logical relations'.

Popper claimed that he had disposed of the inductive method in science for good.[26] Popper's towering influence over 20th philosophy of science has created the impression that inductive practices are negligible events in the history of science, whilst deductive practices are the norm. Popper was much impressed with Einstein's theory of relativity. Here was a good conjecture, freely invented, which resulted in three precise testable predictions (the perihelion advance of Mercury, the red shift of light in gravitational fields and the bending of light near gravitational bodies). But inferential practices are widespread in science and everyday life. In everyday life they take the form of simple, enumerative inductions, which Hume criticized. But in science they are more sophisticated forms of inference, known as abduction or eliminative

---

[26] According to Hacohen (2000: 133–134, 144), he accepted the method of induction in his psychological work until 1929. As he wrote to John Stachel it was not until then that he realized the close link between induction and demarcation.

induction.[27] The inductive props on which our everyday life depends make us believe that we can safely infer from past to future occurrences. Trust is built on such inductive assumptions. Popper does not deny the usefulness of inductive props, even in science. As perceptive students will quickly point out the claim of the universality of laws and theories involves inductive assumptions. This observation is correct. To say, as scientists do, that Newton's law of gravitation applies throughout the universe (as do other laws) is to make an inductive assumption. Life and science would be impossible without inductive practices. There is a distinction between general inductive attitudes in all human affairs and a general inductive method in the acquisition of new knowledge. Hume exposed the logical error in inductive inferences, but he did not deny the practical need for such inferences. Popper, too, accepted this distinction. Not only must there be some room for dogmatism but also for a whiff of verificationism.[28] (*C&R* 1963: 248 [fn31]; *Autobiography* 1974: 40; *Myth* 1994: 94; *Alles Leben* 1996: 121; *All Life* 1999: 41; Keuth 2005: §5.1) A theory needs to be given a chance to prove itself. It should not be discarded too quickly. Popper objected to the inductive method in science on the ground of the asymmetry between verification and falsification. Scientific theories are universal in scope. It is not possible, logically speaking, to arrive at this universality through inductive steps. Einstein agreed.

Popper's anti-inductivism has been much criticized. He did not make his job easier by asserting that he had 'solved' the problem of induction. It is at first hard to understand how all knowledge, including scientific knowledge, should be merely conjectural. Even theories which have withstood severe tests retain their conjectural status. We do not know, says Popper, we only guess. Is it not certain that the earth is a sphere and orbits the sun? Is it not equally certain that atoms and genes exist? It is useful to remember that Popper's theory of rationality is a normative theory: it recommends the search for refutations. (See *Logic* 1959/1980: §§4, 10–11; *Realism/Aim* 1983/1985, Introduction 1982: XXV) Norms express standards of behaviour. But what is the point of norms if nobody behaves in accordance with them? Although Popper's methodology was not meant to be an 'empirical theory', and hence not

---

[27] John Norton, of the University of Pittsburgh, has recently proposed a richly illustrated material theory of induction, according to which inductive inferences (both enumerative and eliminative) are legitimate as long as they occur on a 'case-by-case' basis. Norton (2021: v–viii; 4–8) claims that 'all induction is local' and that 'no universal rules of induction' exist. Particular inferences are warranted by 'background facts in some domain' which 'tell us what are good and bad inductive inferences in that domain'.

[28] Several articles in O'Hear ed. (1995), for instance by Newton-Smith and Lipton, elaborate on these inductive elements. There are, therefore, in Popper's account inductive assumptions. One of the authors who pointed out that 'falsificationism' requires inductive assumptions, was my former colleague Anthony O'Hear (1980). Popper complained to him that he did not like his book, (although he admits that his own account contains a 'whiff of verificationism'). Anthony told me in an email (28/06/20): 'He (Popper) added that I was "product of the modern education"—by which he meant that I was a follower of Moore and Wittgenstein. But perhaps things were not quite as abrasive as it might have appeared at the time (1980). I found out a lot later that he had told a friend of mine that he (the friend) ought to read my book. He (Popper) did not like it, but it was a serious book, or words to that effect'. Miller (1994: Chap. 2) lists a number of such inductive elements and attempts to eliminate them from Popper's account.

'empirically refutable', he still accepted that it could be criticized. (*Parmenides* 1998: 209 [fn46]) So, the accusation has been made that Popper's methodology of rational behaviour does not reflect scientific practice.

> We do have a propensity – an a priori propensity, if you like - to reason "inductively", and the past success of "induction" increases that propensity. (Putnam 1974: 239, cf. Stove 1982; O'Hear 1985; Keuth 2005: §1.122; Norton 2021)

Does it matter whether an inductive or deductive attitude is adopted? It has often been repeated that Popper revolutionized 20th philosophy of science. (Bronowski 1974; Lakatos 1974; Hacohen 2000: Chap. 6) If testability is the core of Popper's fallibilism, what is so special about his approach?

The attitude one adopts towards scientific theories changes the way one deals with them. As Popper never tired of pointing out the inductive method encourages the expectation that the accumulation of evidence makes a theory more probable, eventually acquiring the honorific title 'true'. By contrast in an infinite, open universe, theories remain improbable. (*Realism/Aim* 1983/1985: 252, 256) Popper saw danger in the inductive attitude. Freudians and Marxists tended to consider all evidence as confirmatory. Contrary evidence was dismissed or explained away. The theory could always be saved by ad hocnmanoeuvres. Popper denounced this as an unscientific attitude because it led to dogmatism.[29] In the social world it may lead to intolerance and prejudice. In politics it may encourage totalitarianism. The danger of accepting scientific theories as 'proven' puts them beyond the threshold of doubt. This attitude *discourages* the investigation of alternative theories. Greek geocentrism survived for centuries, despite early indications that it was wrong. The critical attitude towards theories—conjectures and refutations, trial and error—encourages the consideration of alternatives. It encourages the multiplication of competing theories. Popper extended this attitude to a general anti-authoritarianism.

Popper's insistence on severe tests leads to a different attitude towards science. It is not to be treated as an authority, not to be regarded as true or definitive. Scientific theories are tentative. Their acceptance is provisional. And scientists always strive to improve them. Hence they are constantly looking for better theories.

> The more or less universal acceptance of the conjectural character of science has meant a fundamental change in the attitude of scientists towards refutations of mistaken theories, even their own. A similar change in attitude to human fallibility has not yet taken place in politics and within our different institutions. (Popper, *Parmenides* 1998: 65)

Popper now exaggerates in the opposite direction.[30] Some theories are less conjectural than others. Established theories are more reliable than speculative theories. In *Grundprobleme* (1979: 96; 2009: 156) Popper still spoke of the 'graduation in the certainty of our hypotheses'. But the revolution in science, which the General theory

---

[29] In his work on political philosophy he condemned the dogmatism, which he detected at work in Plato, Hegel and Marx.

[30] Popper was prone to exaggerations: induction does not exist, a large part of the knowledge of organisms is inborn, all tests boil down to attempted falsifications or everything is a propensity.

of relativity had triggered, impressed Popper so much that all knowledge claims became equally conjectural.

Not only did Popper regard all knowledge as conjectural, he also objected to calling scientific theories straightforwardly 'true'.[31] Theories only approach the truth even though it may never be known whether they have reached it. It is a regulative idea. (*C&R* 1963: 226, 241; *Objective Knowledge* 1972: 264) If scientific theories are complex structures, then the idea of 'correspondence to the facts' does not really apply to them. Conjectural theories, as he stated in his book on *Parmenides* (1998: 167) are *attempts* to represent physical reality. Many theoretical elements of theories do not correspond to facts: not every property of a theory corresponds to a property of the world. (*Open Universe* 1982/1988: Chap. II, §13, p. 38) Think, for instance, of 'negative velocity', 'negative energy' or 'irrational numbers'. Thus correspondence is a relation between 'symbolic descriptions' and 'physical reality'. (*Myth* 1994: 104) (Box 2)

Box 2: Tarski's Correspondence Theory of Truth

Popper came to embrace Tarski's correspondence theory of truth. (*Objective Knowledge* 1972: 335–340) Tarski's theory was formulated to avoid logical paradoxes, involved in statements like 'I am a liar'. There are good reasons for thinking that scientific theories are not *true per se*, quite independently of Popper's arguments, which were motivated by his fallibilism. Popper himself wrote that every description, including true ones, is selective, 'omitting many aspects of the object described' and augmentative, in that it transcends its evidence by adding a hypothetical dimension'. ('Replies' 1974: 1089) In *Open Universe* (Chap. 2: §13, p. 38) he added that it is not the case that 'for every property of a true theory, there will be a corresponding property of the world'. Theories are layered structures, more so than nets. All theories involve approximations and idealizations. They cannot *literally* correspond to the facts; a better characterization is that theories *fit* the facts. A theory fits the facts, if it satisfies a number of conditions (or constraints), both theoretical and empirical, which lead to its provisional acceptance. That is, a theory is provisionally regarded as 'true' if the conditions it describes are satisfied in the empirical world. Kepler's discovery that planets perform elliptical orbits around the sun is true if these conditions are satisfied in the solar system. Truth is the aim of science—and Popper followed Tarski in describing it as correspondence to facts—but it is really a question of *fit*: the theory satisfies a number of empirical and theoretical constraints. It is mathematically consistent and describes and explains all the evidence in its domain.

---

[31] In his later work he regarded the notion of *verisimilitude* (or *truthlikeness*) as a more realistic aim of science. (*Objective Knowledge* 1972: 57–58) In a panel discussion in the 1980s, he rejected the view, attributed to him, that 'theories are never true'. 'This is nonsense. Scientific theories are the ones, which have survived the elimination process' (*Zukunft* [4]1990: 101; my translation).

3.7 The Method of Abduction

Several roads led Popper to falsifiability. The predictive precision of Einstein's theory stood in sharp contrast with some social science theories (psychoanalysis, Marxism). In the face of universal theories, the asymmetry between verification and falsification had logical weight. His anti-psychologism led to the demand that the 'protocol sentences' of the logical empiricists become objective test statements, with limited generality. But Popper's conflation of 'falsifiability', 'refutability' and 'testability' prompted some scientists to reject what they perceived as 'naïve falsificationism'. I argued that the notion of testability offers a way out of this dilemma. Testability does not just include the requirement that tests be imaginable, even if technically not performable. It also refers to the nature of testing, by direct and indirect means. Direct or indirect testing is a sophisticated form of falsificationism. By indirect testing scientists established the chemical composition of the sun and computed the mass of the earth. Modern cosmologists who investigate the multiverse are particularly interested in this aspect of testing. They add a further feature of indirect testing. The multiverse hypothesis is regarded as the 'best explanation of the observable and experimental data'. (Carroll 2019; Susskind 2006: 300; cf. Vaas 2019) This feature avoids the worry that the multiverse idea is unscientific because it is unfalsifiable. Philosophers refer to this procedure as 'inference to the most plausible explanation'. It is also known as *abduction*, (or 'retroduction') first introduced by the American pragmatist Charles Peirce. (See Freeman/Skolowski 1974: 509–512) Abduction can be thought of as inverse induction. Inductive reasoning, logically unreliable as it is, proceeds from 'some' to 'all' cases or from 'past' to 'future' events. Abduction proceeds from *known* empirical phenomena to the most plausible explanation to account for them. When data are available, science goes in search of an appropriate explanation. This requires the availability of explanatory theories, just as deductivists such as Popper demand. These accounts attempt to fit the evidence.[32] Abductive reasoning is widespread in science where it serves to support a theory, hypothesis or model. Popper himself proposed the above-mentioned *tetradic scheme*, in which tentative solutions are 'measured' against the evidence and eliminated if unfit. Darwin used abduction to defend his theory of evolution against creationism. (Nola/Weinert 2012; cf. Norton 2021: Chap. 9.4) Rutherford employed it to establish, against Thomson's model, that atoms must have a positively charged nucleus, around which negatively charged electrons orbit. (Weinert 2000) There is a body of known facts—from pebbles to planetary orbits—which call for an explanation. Abductive inference infers from the known facts to explanations, which attempt to account for them. Some explanations will be less successful than others. Facts are recalcitrant. The unsuccessful explanations will be eliminated. If it is known that planets perform elliptical orbits, then the stipulation of circular orbits cannot be the right explanation (barring ad hoc evasions.) The hypothesis which best accounts for the phenomena is tentatively retained or 'corroborated' in Popper's parlance. Popper would not call it the most

---

[32] The theories themselves may be generated from conjectures, intuition or inductive generalization.

'likely' explanation. But it is the most plausible relative to the available evidence. It can be overthrown by further evidence.

> (W)e may be entitled to add that the surviving theory is the best theory – and the best tested theory – of which we know. (*Logik* 1935/⁵1973: 373; *Logic* 1959/1980: 419)

This does not prevent the 'most plausible explanation' from making daring predictions. The reader may begin to wonder what all this has to do with Popper. Abduction, naïve or sophisticated falsificationism, retroduction: are we in danger of playing linguistic games? It would be a most un-Popperian game. However, a closer look reveals that sophisticated falsificationism is really a form of abduction. First add to the 'body of facts' the existence of problems. It was one of Popper's central and enduring insights that all organisms are engaged in problem-solving. Problems invite tentative solutions, error elimination and give rise to the formulation of new problems. Kepler faced the problem of explaining the cause of the elliptical orbits he had discovered. Astronomy had to wait for Newton's laws to find an answer. Darwin faced the problem of the age of the earth. In Darwin's days William Thomson (Lord Kelvin) estimated it to be no more than 100 million years. Such a time span would have been far too short for natural selection to do its work. The Darwinians had to await the discovery of radioactive decay (1901) to learn that the age of the earth numbered in hundreds of millions of years.

Popper was aware of abduction. He agreed that Peirce partly anticipated his 'fallibilism'. But he regarded his own epistemological model as more 'radical' than Peirce's.

> Peirce wrote before Einstein shattered our belief in Newton's most wonderful and most successful theory. (…) My more far-reaching fallibilism, on the other hand, is the direct result of Einstein's revolution. ('Replies' 1974: 1065)

Francis Bacon (1561–1626), English philosopher and statesman. *Source* Wikimedia Commons

There is in fact another predecessor of the method of 'abduction'. Francis Bacon is often associated with the inductive method, which Popper rejected. It is the inference from particular to general cases; also known as ampliative or enumerative induction. Bacon's ideas, however, were more sophisticated than the caricature to which Popper contributed. To be fair to Popper, his attitude to Bacon, whose ideas he often discussed, was ambivalent. He did not think much of him as a philosopher of science but conceded that he was the 'spiritual father of modern science'. (*Myth* 1994: 195) He did not appreciate Bacon's method of science because he associated it with naïve empiricism. For Bacon, 'all induction consists in the collection and (statistical) tabulation of instances, especially confirming instances'. (*Realism/Aim* 1983/1985: 256) But confirming instances, as he had said in his criticism of Adler and Freud, were easily obtained. They could not form the basis of generalizations. Enumerative induction therefore failed as a method of science. The Baconian myth, according to Popper, is that 'observation is the *true source* of our scientific knowledge'. (*C&R* 1983: 138; italics in original; cf. *Zukunft* [4]1990: 66; *Autobiography* 1974: 62) As Popper's supporter Peter Medawar pointed out, induction by enumeration stresses the primacy of facts, is preoccupied with justification and has no clear sense of the critical function of experiments. (Medawar 1984: 85–103)

But at times Popper realized that he was painting an oversimplified picture of Bacon's method. In his notes on corroboration, originally published in the *British Journal for the Philosophy of Science*[33] (1954–1958), he admitted that Bacon and other inductivists were aware of the method of eliminative induction. He describes this method as 'induction by the method of eliminating or refuting false theories'. (*Myth* 1994: 105) He rejects the comparison with his own method of Critical Rationalism. The purpose of elimination was to find the 'true' or 'most probable theory'. According to Popper we should, however, hold on to the riskiest and most severely tested theory. (*Logic* 1959/1980: 419; *Logik* 1935/[5]1973: 373)

It is indeed the case that Bacon rejected the simple method of enumeration as 'puerile'. His preferred method was the method of elimination and exclusion.

> The induction which proceeds by simple enumeration is puerile, leads to uncertain conclusions, and is exposed to danger from one contradictory instance, deciding generally from too small a number of facts, and those only the most obvious. But a really useful induction for the discovery and demonstration of the arts and sciences, should separate nature by proper rejections and exclusions, and then conclude for the affirmative, after collecting a sufficient number of negatives. (*Novum Organum* Book I, §105; cf. §125, Book II, §§19, 36)

He contrasted his method with that of the empiricists and rationalists.

> Those who have treated of the sciences have been either empirics or dogmatical. The former like ants only heap up and use their store, the latter like spiders spin out their own webs. The bee, a mean between both, extracts matter from the flowers of the garden and the field, but works and fashions it by its own efforts. The true labour of philosophy resembles hers, for it neither relies entirely or principally on the powers of the mind, nor yet lays up in the memory the matter afforded by the experiments of natural history and mechanics in its raw state, but changes and works it in the understanding. We have good reason, therefore, to derive hope from a closer and purer alliance of these faculties (the experimental and rational) than has yet been attempted. (*Novum Organum* Book I, §95)

---

[33] Now Appendix *ix of his *Logic of Scientific Discovery*. Popper (*Myth* 1994: 86–87) acknowledges that Bacon was aware of the defect of simple induction by enumeration.

Bacon proposed a more sophisticated method: researchers should look for the best or most likely explanation. He even suggested the use of crucial experiments to choose from amongst competing theories. (Book II, §36) Is Bacon's more sophisticated method of ampliative induction so different from Popper's approach? It is true that Bacon, like the seventeenth century generally, sought the certainty of knowledge. He aimed at the most probable rather than the riskiest explanation. In this respect there is a difference between Bacon and Popper.

But Popper was not the first to reject enumerative induction. In *Logic* (1959/1980: Chap. I.1. p. 30 [fn 5]) he refers to his friend Viktor Kraft and to Justus von Liebig, who is regarded as the founder of organic chemistry. But there were others, such as the French physiologist Claude Bernard and his compatriot Henri Poincaré, or the English polymath William Whewell, who stressed the importance of hypotheses guiding experimental research. (Bernard 1865; Poincaré 1968: Ch. IX) In contrast to Bacon, Liebig emphasized that research is 'deductive and a priori'. The formulation of a hypothesis necessarily precedes the experiment.

> An experiment not preceded by a theory – that is, by an idea – stands in the same relation to physical investigation as a child's rattle to music. (Liebig 1863: 263)

Liebig shared the misunderstanding of Bacon as a simple inductivist. Bacon did not include problems in his list of facts. Popper's stunning insight was that science, even politics, starts and ends with problems. It is a never-ending march towards ever more complicated problems and tentative solutions. Now if 'body of facts' is replaced by 'body of problems' what emerges is Popper's tetradic scheme. (*Autobiography* 1974: §29) The problems can be practical or theoretical. The most likely explanation – on Bacon's scheme - is the one that survives the error elimination. What survives on Popper's scheme are unrefuted hypotheses and the most conjectural explanations, because they are at the greatest risk of elimination. It leads to the emergence of new problems ('what is gravity?') and theories (the curvature of four-dimensional space–time) which may eventually face defeat themselves.

Popper was right that terminology should not dominate the discussion. But if one adopts this broader perspective, one begins to see that Carroll had a point. Testability, abduction, sophisticated falsificationism—it captures the right way to think about science. A further difference between Bacon and Popper is that

> (h)e was the first to stress that the hypothetical character of a theory is never mitigated, however numerous and however breath-taking its empirical successes (…).

Popper grants evidence no positive role 'in the assessment of theories'.

> Indeed, it is not so much inductivism that he demolished, as justificationism; the view that the rationality of science (…) consists in the pursuit and attainment of (wholly or partially) justified truth, rather than of truth. (Miller 1997: 376)

Falsification can be interpreted as a limiting case of abduction. Abduction and testability are contrastive procedures. The same evidence is used to assess respective values of several competing theories. In his later work, Popper imposed some stringent constraints on the acceptance of new theories, which are compatible with the idea of testability. It is revealing that in his *Autobiography* (1974: §10), Popper summed up his *Logik* as containing a trial-and-error theory of the growth of knowledge or more precisely 'through the elimination of errors'. He felt that he had eliminated the errors of the *Vienna Circle*. He was not intimidated by the inductivist traditions in science nor by the authority of classical physics. He insisted on objectivity and rational criticism.

3.8   Emigration

In a letter from 1936 Tarski informed Popper that his *Logik der Forschung* was much discussed in Poland [354: 08], so it was in England. But no lectureship had materialized in the USA or the UK. The Faculty of Moral Science at the University of Cambridge offered him academic hospitality for a year, at £150. Popper declined the Cambridge offer in favour of Waismann. But he had now written an influential book on the theory of scientific knowledge, which finally resulted in a job offer. On October 25, 1936 Popper applied both for a lectureship and a professorship at the Department of Philosophy and Education at Canterbury University College in Christchurch on the South Island of New Zealand. His application stated G. E. Moore, Professor of Philosophy and J. H. Woodger, Reader in Biology, as referees. He also submitted testimonials from the psychologist Karl Bühler (23/10/36), the philosophers Rudolf Carnap (02/10/36), Susan Stebbing (12/10/36), Bertrand Russell (12/10/36)—who praised him as 'a man of great ability'—and Alfred Tarski (mentioned above). The physicist Niels Bohr (16/10/36) expressed his 'high appreciation of the scientific and personal qualifications of Dr. Karl Popper as a philosopher especially interested in the rational foundation of science' ([Hansen 3.2; Testimonials 406.1-5]; Hacohen 2000: 323). The references highlighted the achievement of *Logik der Forschung*, which according to Carnap was 'one of the most valuable books in epistemology of recent years'. It was the falsifiability principle that got him his job. He had to make do with the lectureship for the professorship went to the anthropologist Ivan Sutherland, with whom Popper would have a serious fall-out. At least it was a permanent job at £400. (Watkins 1997b: 652; Brudny 2002: 106–107)

The Poppers left a continent over which the dark clouds of fascism were hanging. Popper had seen signs of the looming danger already in 1927: anti-Semitism, the rise of totalitarianism, and Hitler's aggressive policies. The Nazi occupation of Austria brought fascism to his homeland. In New Zealand Popper would turn to political philosophy. He regarded these writings as *his* contribution to the war effort, his contribution to the fight against Nazism. Popper's anti-authoritarianism would not be aimed at scientific traditions but political rule. It resulted in the publication of the *Open Society and its Enemies* and *The Poverty of Historicism*.

Chapter 4
Popper's Life in New Zealand

The appeal of Utopianism arises from the failure to realize that we cannot make heaven on earth. (K. Popper, *Conjectures and Refutations* 1963: 362)

Karl and Henny Popper in Christchurch in early 1945.

Karl and Hennie in Christchurch, New Zealand, 1945. *Source: NZ Science Review* **46** (1989)

The Poppers arrived in Christchurch in the first week of March 1937, just in time before the beginning of the new academic year. (Fig. 4.1) (March is the beginning of autumn in the southern hemisphere.) Popper was the only full-time philosophy tutor at that time. He had a heavy teaching load: Ethics, History of Philosophy and Introduction to Philosophy, Institutional Ethics and Politics, Logic, Philosophy of Nature and of Society and Scientific Method (not all these courses were taught every year) [300.02]. Although his time in New Zealand was not without difficulty (see below), his prodigious capacity for work was undiminished. Apart from some technical work on probability and logic, his time in the southern hemisphere is noteworthy for his contributions to political philosophy and his turn to the methodology of the social

© Springer Nature Switzerland AG 2022
F. Weinert, *Karl Popper*, Springer Biographies,
https://doi.org/10.1007/978-3-031-15424-9_4

sciences. The step from the philosophy of the natural sciences—especially physics—
to the methods of the social sciences—especially sociology—should not come as a
total surprise. The events, in *Hörlgasse* (1919), had led him to a critical assessment
of Marxism. Two years later, in 1921, he abandoned it. He had studied psychology at
the Pedagogic Institute, had reflected on the methodology of psychology in his Ph.D.
and had given a lecture on the 'poverty of historicism' at the *LSE* (1936). A year
after Popper's arrival in New Zealand, Hitler's army marched into Austria. Popper
had been reluctant to publish his critique of Marxism. But the *Anschluß* (12 March
1938) broke his reluctance. His emigration re-awoke in him an interest in political
philosophy.

> I also felt that I could no longer hold back whatever knowledge of political problems I had
> acquired since 1919; I decided to put "The Poverty of Historicism" in a publishable form.
> What came out of it were two more or less complementary pieces: *The Poverty of Historicism*
> and *The Open Society and Its Enemies* (which at first I intended to call: "False Prophets:
> Plato – Hegel – Marx"). (*Autobiography* 1974: 90; italics in original)

In fact, for a long period Popper could not decide on the title of *Open Society*.
Intense discussions between Popper, Gombrich and von Hayek took place before it
was agreed that Routledge would publish the book under its familiar title. Although
he wanted to fight against totalitarianism, he confessed to Woodger that he would
never have written *Open Society* 'but for the war' (02/07/45) [262.17]). The two
'pieces' became separate books, published at the end of his New Zealand period.
Poverty is a contribution to the philosophy of the social sciences. It considers how
the methods of the social sciences compare with those of the natural sciences. *Open
Society* is a critical study of the political ideas of Plato, Hegel and Marx and their
methodology. He treats Aristotle as a minor figure: 'Aristotle's thought is entirely
dominated by Plato'. (*Open Society* II: 2) (He later admitted that he had been unfair
to Aristotle and misinterpreted him 'many times', *Parmenides* 1998: 275). He is
openly hostile to Plato and contemptuous of Hegel. He is critical of Marx's general
approach but sympathetic to some aspects of his work. It is not important for present
purposes to decide whether Popper's assessment of Plato and Hegel is technically
correct. Historians of philosophy have been critical of Popper's verdict. What is
important, however, is Popper's resolute defence of the *open society*. Why did he
single out Plato, Hegel and Marx as its enemies? Popper distinguished the *open* from
the *closed* society. An open society is one, which values the freedom of the individual
(over the collective), prefers piecemeal social reform (over violent revolutionary
change), embraces democracy as a political system and allows the operation of the
free market, within certain limits. Popper emphasized, in strong agreement with Kant,
the virtues of cosmopolitanism. He hated all forms of nationalism, including Jewish
nationalism (letter to Woodger 31/03/78 [568.03]). This becomes abundantly clear
in *Open Society* (I: Chap. 9, note 7(1); II: 49). As we shall see, Popper remained
committed to his defence of the open society. In his later reflections, the list of
priorities in an open society shifts to the promotion of peace, halting the population explosion, proper education for children and censorship of the media if necessary. (*Lesson* 1997: 34f) An article in *The Economist* (April 23, 1988) considers

electoral systems. A *closed* society is mired in a rigid worldview. It is hostile to internal criticism and follows tradition blindly. It does not provide the individual with the essential freedom that is even more important than economic success. (*Ich weiß* [2] 1992: 68)

The notion of an open society struck a note with many public figures. The former German Chancellor Helmut Schmidt, a social democrat, was impressed with Popper's rejection of Marxism and his plea for piecemeal reform. His emphasis on 'trial and error', introduced into politics, highlights the capacity of democracy to limit the damage, which unfit rulers can inflict on society. (Schmidt 1987) In interviews he credited Popper with having guided him in his political affairs. As evidence he sent Popper a copy of his last speech in the budget debate before the Bundestag (10/09/86). It reflects on the monetary and foreign policies of the Kohl government. But then it becomes more philosophical: he underlines the role of the constitution and the rule of law, the rights of the individual in a democracy, the importance of social justice, peace and respect for the opponent, the need for compromise and piecemeal engineering in a pluralistic state. The edifice of society must be built on reason and realism. Idealism is also necessary but it must be subject to critical scrutiny. He finishes by declaring that 'no enthusiasm should be greater than the sober passion for practical reason' [562.01; my translation]. Popper was touched and promised him (09/89) a copy of a short address he gave to a group of members of the Social Democratic Party in the House of Lords [562.01]. Helmut Kohl, also a former Chancellor and Christian Democrat, sent Popper a telegram on occasion of his 75th birthday, emphasizing his contributions to the understanding of democracy [316.01]. Kohl also wrote the Preface to the Routledge Classics edition of Popper's Autobiography *Unended Quest*.[1] He praised Popper 'as one of the most significant champions of the open society' and for having destroyed the fallacy of the 'belief in the historic mission of one race or class'. In Britain the Labour politicians Anthony Crosland and Denis Healey were, according to Bryan Magee (letter dated 17/09/76 [323.01]), influenced by Popper; so was the Conservative party politician and Vice Chancellor of Leeds University, Edward Boyle. But Magee (who in 1982 switched from Labour, which he saw under threat, from the Left to the Social Democrats) was the most loyal and influential Popperian.[2]

---

[1] It appeared in 2002 but Fontana had published a revised edition of *Autobiography* in 1976, which a year later had sold 14,000 copies [355.08] and by 1979 28,000 copies [272.08.] It had received some good reviews [334.04] but not yet in the *Times Literary Supplement* so that Popper suggested it to them for a review (09/04/77) [355.07].

[2] Political programmes like the open society and piecemeal engineering also appealed to Thatcherite politicians, as Peter Munz pointed out in his *Obituary*. Apparently Keith Joseph, a politician who served under four Conservative prime ministers, told Margaret Thatcher that 'she was a Popperian', as John Worrall mentioned in our interview. Magee (1995: 266–267) argued that Popper's approach can be adopted by all who are committed to democratic politics. But it 'sits more comfortably with a left-of-centre position (…) because it gives rise to a radical attitude towards institutions.' Magee revealed to Popper (02/08/92) that Mrs. Thatcher never talked to him about Popper's work. But she talked to Magee about his book on Popper when she was 'an obscure Opposition MP' [550.02]. Parvin (2013: 98) also notes that Popper's views have influenced liberal, conservative and social democratic thought alike. But his insistence on individualism and gradual reform had endeared him

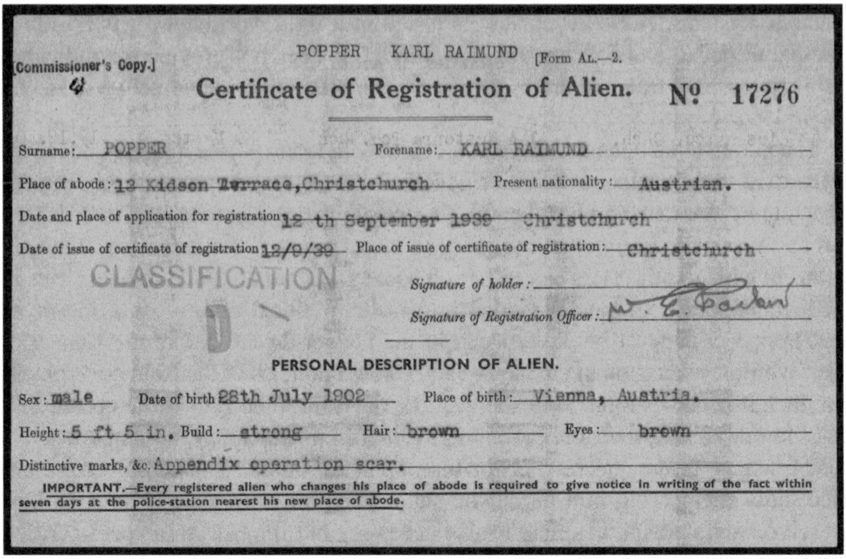

[Commissioner's Copy.] POPPER KARL RAIMUND [Form AL.—2.

4

Certificate of Registration of Alien. No 17276

Surname: POPPER Forename: KARL RAIMUND

Place of abode: 12 Kidson Terrace, Christchurch Present nationality: Austrian.

Date and place of application for registration 12 th September 1939 Christchurch

Date of issue of certificate of registration 12/9/39 Place of issue of certificate of registration: Christchurch

CLASSIFICATION D

Signature of holder:

Signature of Registration Officer:

PERSONAL DESCRIPTION OF ALIEN.

Sex: male Date of birth 28th July 1902 Place of birth: Vienna, Austria.

Height: 5 ft 5 in. Build: strong Hair: brown Eyes: brown

Distinctive marks, &c. Appendix operation scar.

IMPORTANT.—Every registered alien who changes his place of abode is required to give notice in writing of the fact within seven days at the police-station nearest his new place of abode.

Fig. 4.1 Popper's certificate of registration of alien, issued on September 12, 1939, in Christchurch, New Zealand. *Source* Wikimedia Commons

Magee's adherence to Popper's philosophy is well documented in their lengthy, handwritten correspondence, in his books, interviews and TV programmes. Much travelled Mr. Magee regarded Popper as the 'greatest philosopher since Kant' (22/08/69) [322.18] and a 'man of genius' (03/10/70) [549.19]. Magee was many things in his life: a British Labour MP, a broadcaster, music lover, novelist, politician, theatre critic and an academic philosopher to boot. He was on good terms with many of the leading intellectuals of his day. No one did more to popularize Popper than Magee. He was eager to tell his mentor (20/09/83) that the influential Oxford philosopher Peter F. Strawson praised him as 'the best serious popularizer of philosophy for half a century … in every medium' [323.04].

Magee was a 28-year-old parliamentary candidate when he wrote to Popper (14/10/58) that he had been deeply influenced by *The Open Society*. He wished to meet its author—which he did in early 1959—because he was interested in the practical applications of Popper's views. He wanted to discuss their effect in public life. He felt that there was a need to convince politicians, rather than dons. At the beginning of 1959 (03/01) he sent Popper a nine-page letter to voice his disagreement on a number of points that Popper had raised in his lecture 'The History of our Time', (now Chapter 19 of *Conjectures and Refutations*). In his handwritten,

---

to conservatives (p. 112). Hacohen (2000: 449) and Shearmur (1996: 53) agree with Magee that the *Open Society* is more in line with liberal and social democratic thinking. One should not forget that Popper accepted the need for political interventions in a welfare state and remained committed to 'negative utilitarianism', i.e. the view that the state should minimize suffering and protect the weakest members of society.

twenty-three page reply Popper graciously accepted his 'valuable criticism' and later (01/02/59) promised that he would amend his paper in the light of Magee's criticism. He also told Magee to 'drop the title "Professor"'. Thereupon a lifelong friendship and co-operation developed between the two men [322.17].

Popper's crusade against closed societies inspired George Soros, the American-Hungarian international financier, to create his *Open Society Fund* (1982). Soros first contacted Popper in 1952 because *The Open Society* had impressed him. In the ensuing years he sent Popper manuscripts for comments and feedback. The purpose of the *Open Society Fund* was the promotion of 'an open society and a critical mode of thinking, which is indispensable to it' (letter to Popper (28/09/87) [563.15]). In 1991 Soros established the Central European University in Budapest, Prague and Warsaw. Popper was intrigued and asked him to send more information [563.15]. (Miller 1997: 400) Soros, who, in February 1962, attended Popper's seminar at the *LSE*, condemned the 'cult of success, which undermines the open society'. 'We need a commitment to the open society', he wrote, 'because it is the right form of social organization'. It provided a framework and aimed 'to open up closed societies'. Yet it was only a necessary, not a sufficient condition for freedom and prosperity. Whilst Popper identified totalitarianism as the main enemy of the open society, Soros (1997) argued that it had been replaced by *laissez-faire* capitalism, of which Popper never approved.

> I believe that the injustice and inhumanity of the unrestrained 'capitalist system' described by Marx cannot be questioned.... (*Open Society* II: 124)

This quote anticipates a theme, to which Popper returned in his 'political' phase: it is the need for state intervention. He pleaded for a mini-state as a regulative principle; a state that is neither paternalistic nor *laissez-faire*; and a state that respects the rule of law.

> We need freedom to prevent the state from abusing its power, and we need the state to prevent the abuse of freedom. (*Lesson* 1997: 73; *Alles Leben* 1996: 227)

His objections against the three 'prophets' revolved around their plea for a paternalistic state.

4.1 Plato

For readers who are not familiar with Plato's political philosophy, here is a brief résumé. Plato constructed a model of the ideal state. It is characterized by the strict separation of classes. Philosophers form the head of government. The guardians are its executive arm. Merchants, tradespeople and workers constitute the third class. Not only is the ideal Platonic state strictly hierarchical, its hierarchy is justified due to the different virtues, which each class possesses. Philosophers must be 'philosopher kings' because they possess a special insight into what constitutes good government. Good governance requires wisdom, character virtues and a privileged understanding of justice on the part of the rulers. Whilst the rulers possess these timeless character-istics, they are deprived of material possessions and family life. They must govern for the good of the whole society. Plato's reasoning relies on an all too familiar fallacy (the fallacy of composition): a person who is good at one thing must be good at many other things. But good philosophers do not necessarily make good politicians. (Einstein was wise enough to turn down the offer of the presidency of Israel.) The guardians (or warrior class) have the virtue of courage. They carry out the political orders of the philosopher kings. Finally, ordinary citizens pursue their diverse activ-ities, but do not (normally) qualify for higher office. Plato expressly warns against a mixture of the social classes. It will lead to chaos and instability.

Two features of this conception of the ideal state stand out. First, Plato follows the Greek tradition of thinking that people are *born* with certain characteristics, rather than acquiring them. It was natural to the Greeks to think that some are born to rule and others to obey. *Our* way of thinking marks a significant departure from this Greek tradition. Plato even invents the myth of the metals. Rulers are born with gold in their souls and guardians with silver, while merchants and craftsmen have a bronze soul. Plato holds that it is legitimate to spread this myth in order to promote social harmony. It is a virtuous lie but a propaganda tool in Popper's eyes. The problem is of course the appeal to the inborn traits of people's characters. Your birth fixes your character. The second feature, related to the first, is Plato's theory of justice. It follows from his view of the unchangeability of character. A society is just when every member fulfils their function, according to their 'nature'.

Plato also conceives of a ranking of governments, in descending order: the best government is his ideal state, or aristocracy (rule by wisdom), followed by timocracy (rule by honour and duty), oligarchy (rule by wealth), democracy (rule by the many) and tyranny (rule by fear). He dismisses democracy as an anarchic type of society, in which everybody is equal, and the state ship is commanded by incompetent captains. It is important to note that tyranny appears at the bottom of the hierarchy. It is based on injustice and arbitrary rule. Societies degrade in this cascading manner because of internal strife or the lowering of standards.

I have mentioned two fundamental misconceptions in Plato's ideal state. Given Popper's 'socialist' credentials,[3] it may come as no surprise that he rejected Plato's model state. He devotes almost the whole of the first volume of *Open Society* to a critical evaluation of Plato's political philosophy. (Aristotle's political philosophy, as mentioned, is treated as totally derivative of Plato's.) Popper accuses Plato of devising a recipe for a closed society. The problem, as Popper sees it, is the starting point: the ideal, which according to him is located in a mythical past. Compared to the ideal every other type of state looks degraded and decayed. It has moved away from the past ideal. Plato is hostile to change because it brings about the degradation of society.

---

[3] In an exchange with Jacques Monod (24/12/71) he confessed that he was 'emotionally a socialist' in a humanist, not a political sense. He was in favour of a more 'fraternal society' but against the nationalization of 'the means of production', which would lead to dictatorship and nationalism [329.23].

There is a more Weberian way of looking at Plato's ideal state. It does not hark back to an idyllic past but projects an *ideal type*. It presents a yardstick, whose purpose is to measure the deviation of actual communities, of existing societies, from the ideal type. As we shall see, Popper comes close to Weber's methodology of ideal types in his *The Poverty of Historicism*. Weber defined the salient features of ideal types: economic, social and political systems (capitalism, Protestantism, democracy), of historical periods (feudalism) to compare them to their actual counterparts. For instance, freedom of movement and opinion, of religious worship are salient features of a democratic society. The question for the social scientist is to which extent actual democracies depart from these features. Weber's ideal types are conceptual models towards which actual entities strive, whereas Popper interprets Plato's ideal state as a real, past state from which actual societies have digressed. Even though Plato condemns tyranny as the lowest form of government—its most decayed form, one of arbitrary power—Popper objects to several features of Plato's ideal state. They stand in stark contrast to his model of an open society.

- Plato's closed society lacks a clear distinction between natural laws and social conventions. Social or normative 'laws' can be changed by human intervention, unlike the laws of nature. Humans cannot change the orbits of the planets, but they can change the norms and values which govern social behaviour. Plato treats human characteristics (being a slave) and social institutions (slavery) as facts of nature. Humans are 'by nature' born with gold, silver or bronze in their souls. Popper rejects the preoccupation with the 'nature' or 'essence' of things, which he also found objectionable in Aristotle.
- Plato's ideal submits the individual to the interests of the state. The collectivity— the state—is greater than the sum of individuals. Rulers are preoccupied with the health of the state, which they equate with justice. Popper contrasts this collectivism (also known as political 'holism') with his own emphasis on individualism. It requires the individual to be protected from arbitrary state interference. After the events in *Hörlgasse*, Popper objected strongly to the adherence to the party line. Both the contrasts between individualism and holism as well as that between natural laws and social conventions play a major role in his philosophy of the social sciences.

It is debatable whether in fact Plato's ideal state is totalitarian in nature. The Philosopher Kings live a Spartan lifestyle; very talented individuals can rise to their rank; and they govern strictly for the benefit of all citizens. (Wild 1974) As we shall see, the holism/individualism issue is an ongoing debate in the philosophy of the social sciences. But Popper's objection to the conflation of laws of nature and social convention is an important point. It informs his views on the social sciences. Popper's next two objections have equal weight.

- He raises the question of leadership, to which he would often return. Plato, he points out, asks the question of 'who should rule'? His answer is: the philosophers, because they alone possess a proper understanding of what constitutes a 'good' society. And they possess the wisdom and the qualities to rule. According to Popper it has been the predominant question in Western political thought. Marx, for instance, held that the working class should rule. Popper holds that it is the wrong question. One should not ask 'Who should rule?' since it invites an authoritarian answer. (In a similar way, one should not ask 'what are the best sources of knowledge?' See *C&R* 1963: 25) One should ask instead: 'How can we rid ourselves of poor rulers, without bloodshed?' (And: 'how can we eliminate error?') He rightly points out that Plato ignores the question of the institutional control of rulers. (*Open Society* I: Chap. 7, §3) Popper draws a fundamental distinction between persons and institutions and between office holders and offices. What protects citizens against the abuse of power—dictatorship, tyranny—are the right constitutions, institutions and the rule of law. (*Open Society* I: Chap. 7, §2) The question is, of course, what are the right constitutions, institutions and laws? Popper's answer is democracy and the open society. In this respect, as in others, he is a true child of the Enlightenment. It reveals, in my assessment, sympathy for republicanism.

- One final objection is of great significance, although it affects Hegel and Marx more than Plato. It is the rejection of what Popper calls 'historicism'.[4] This is the doctrine that the evolution of societies is subject to strict laws in a way similar to physical systems. Physical systems obey physical laws, and social systems obey social laws. Hegel and Marx were by no means the only nineteenth-century thinkers who held this view. John Stuart Mill, August Comte, Oswald Spengler and Arnold J. Toynbee are some of the names associated with the belief that societies evolve according to fixed patterns. It is almost a natural consequence of the Scientific Revolution and the Enlightenment.

The Scientific Revolution (1543–1687) convinced scientists that the universe could be explained rationally because its workings were subject to mathematical rigour. Long before Newton's laws of motion, Kepler had formulated his three laws of planetary motion. The Enlightenment philosophers (ça 1600–1800) argued: if mathematics can help to comprehend nature, the social sciences can help to understand society rationally. Societies do not obey natural laws but societal laws. Consequently, Comte argued that all societies necessarily evolve according to three successive stages: from a theological to a metaphysical stage and finally a positive stage. The latter put the emphasis on facts, whilst the former stages stressed religion and

---

[4] This term should not be confused with 'historism', as Popper points out. (*Open Society* II, 208; *Poverty* 1960: 17) Historism is a philosophical and historiographical theory. It originated in nineteenth-century Germany (under the term *Historismus*). It became influential in nineteenth- and twentieth-century discussions of the status of the social sciences *vis-à-vis* the natural sciences. It was defended by historians and philosophers, such as Leopold von Ranke, Johann Gustav Droysen, Wilhlem Dilthey and Friederich Meinecke. Historism refers to the historical relativity of culture and society. In Dilthey's version it led to a new understanding of the social sciences, in opposition to the French School of Henri de Saint Simon and Auguste Comte. The French School of sociology conceived of the social sciences as social physics, based on the notion of explanation. By contrast, Dilthey placed the social sciences near the discipline of history, based on the notion of understanding.

abstract ideas, respectively. Marx also believed that he had detected a necessary sequence of historical stages of society before it could reach socialism.

Popper accuses Plato of promoting a 'historicist theory of society'. His 'historicist sociology' teaches that 'existing states were decaying copies of an unchanging form or idea'. (*Open Society* I: Chap. 4, §5) The cause of social deterioration is strife amongst the ruling elite. This accusation can only be upheld on the basis of Popper's interpretation of a past ideal state, rather than an ideal type, from which actual societies deviate. It is not important in this context to decide whether Popper's interpretation is contrived or mistaken. What is important is Popper's question: Do societies evolve according to societal laws? This question became more prevalent in the Enlightenment period. The laws of development, which Hegel, Marx, Comte, Spengler and others believed to have identified, are not subject to human negotiation. On this issue, Popper shed new light. A consideration of this question with respect to Hegel and Marx will allow us to make a transition to Popper's *The Poverty of Historicism*. This book contains influential reflections on the nature of the social sciences.

4.2   Georg Wilhelm Friedrich Hegel

Georg Wilhelm Friedrich Hegel (1770–1831). *Source* Wikimedia Commons

Out of the three 'prophets' Popper reserves the harshest judgement for Hegel. Despite his reservations about Plato's political views, he considered him to be a great philosopher. He wrote to Gombrich (23/08/43) that Hegel never intended to be understood; 'he wanted to impress, to confuse, to dazzle his readers [300.03]. [5]Hegel

---

[5] Needless to stress that more positive assessments of Hegel's philosophy exist: for instance Herbert Marcuse (1954) and Charles Taylor (1975). The historian Malachi Hacohen (2000: 402), who is generally sympathetic to Popper's political philosophy, brands his portrayal of Plato and Hegel as 'sophisticated anachronism'. But Schopenhauer also ridiculed Hegel as a charlatan and his

was dishonest. There is only one good idea in Hegel's philosophy: 'the indebtedness of reason to tradition'. (*Open Society* I: 59) Popper had a visceral dislike of 'verbiage', of which he believed to find a rich harvest in Hegel's writings. For the same reason, Popper never found the patience to discuss Heidegger or even Wittgenstein. They were too arrogant he told P. A. Schilpp (21/09/69) [334.02]. And his debates with members of the Frankfurt School of Critical Theory (Theodor Adorno, Max Horkheimer, Jürgen Habermas) in the 1960s reduced mainly to accusations of vacuous thinking under the disguise of bombastic language. He reproaches Hegel with worshipping the Prussian state. He casts him as 'the missing link' between Plato and modern totalitarianism. (*Open Society* II: 28–29; 60–78; cf. Hacohen 2000: 432; Keuth 2005: §11.3; Parvin 2013: 72–73) It is difficult to defend such a thesis. Popper did believe that ideas can exercise an effect on human action. But he rejected the nineteenth-century view that the history of society passes through law-like stages—be it towards Comte's positive age or Hegel's Absolute Spirit. In other words, history is contingent. It is dependent on unpredictable human agency and accidental events. (Helmut Kohl, the former Chancellor of Germany, emphasized this aspect of Popper's philosophy in some of his speeches.) Between the publication of Hegel's *Phenomenology of Spirit* (1807) and Hitler's totalitarianism (1933–1945) so many contingent events intervened that the Darwinian allusion to a 'missing link' is misplaced. Hitler's *Mein Kampf* makes no references to Hegel. A dictator needs no inspiration either from Plato or Hegel. A missing link, in Darwin's work, is an evidential gap in the history of a species from its earliest ancestors to its present population. Modern totalitarianism is not a descendent of the Prussian state.

The impact of human agency and the presence of trends, rather than genuine laws, in the social world make the historical contingency of human affairs practically unavoidable. It was already emphasized by Enlightenment philosophers such as the Italian political theorist and historian Giambattista Vico. Popper mentions the German sociologist Max Weber only occasionally in his writings. But there is evidence that he shared Weber's insight: that ideas do not determine but instead *channel* social action. He repeatedly stressed that ideas play an important role in history. (*C&R* 1963: 332, 373; *Auf der Suche*[2] 1987: 73, 107, 155, 200; *In Search* 1992: 59, 90, 178)

Hegel's philosophy is all about the power of the mind. He sees history as a movement towards the realization of the Absolute Spirit, towards the ideal, which he identifies with the Prussian state. It is the dialectical deployment of reason, which Hegel identifies with the real. The most famous example of Hegel's method of dialectics, of thesis and anti-thesis culminating in synthesis, is illustrated in the master–slave (lord and bondsman) relationship (*Phenomenology of Spirit* 1807: Chap. IV). An understanding of this relationship is important for an appreciation of Marx's historical materialism. At first the master is everything and the slave is nothing. The slave works whilst his master remains idle. Through his work the slave transforms nature

---

disrespect for the professional philosophers of his day matches Popper's irreverence towards experts. Schopenhauer's dismissive attitude left its mark on Popper's discussion of Hegel. It is urgent, he writes, 'to continue Schopenhauer's fight against this shallow cant'. (*Open Society* II: 79)

and society. (One of Hegel's abiding insights, which Popper acknowledges, was that 'we are what we are through the work of history.') The slave gains mastery of the world around him. Whereas at first he is dependent on the master, the master now becomes dependent on the slave. The slave gains freedom, which the master loses. This opposition of thesis and anti-thesis must be, according to Hegel, resolved in a new synthesis. The master and the slave must reach a state of mutual respect and equal rights.

Hegel's dialectical method therefore consists of the overcoming of the opposition (thesis and anti-thesis) towards a new level: synthesis. In a famous article, 'What is Dialectic?' (1940; reprinted in *C&R* 1963: Chap. 15) Popper launched a sustained attack on Hegel's method. It leads to the acceptance of contradictions. It cannot distinguish true from false ideas. Logic shows that any statement can be derived from a contradiction. Science, however, aims at the elimination of contradictions. Science works by trial and error, not by dialectical spirals. The trial-and-error method: learning from mistakes is the method of 'all empirical sciences'. Science also distinguishes between the mind and the world: this thought expresses Popper's common-sense notion of realism in a nutshell. Hegel, however, confused his thesis that 'the world is mind-like' with the Kantian doctrine that the 'world is described as mind-like'. The trial-and-error method, which Popper introduced in this article, also serves as a 'rational method of piecemeal engineering' in society. (*Open Society* I: 167)

4.3 Karl Marx

Karl Marx (1818–1883). *Source* Wikimedia Commons

Popper's assessment of Marx is far more balanced and unbiased than his discussion of Plato and Hegel. He pays tribute to Marx's humanitarian instincts, counting him amongst the 'liberators of mankind'. (*Open Society* II: Chap. 17, p. 122) He agrees with Marx that philosophers should help to change the world. (*C&R* 1963: 337)

But Marxism is the 'most dangerous form of historicism'. (*Open Society* II: Chap. 13, p. 81) Marx confused small-scale scientific prediction with large-scale historical prophecy. Popper grants that Marxism was at first a scientific theory, because it made predictions, for instance the downfall of capitalism and its replacement by socialism. But when these predictions turned out to be mistaken, it shielded itself against falsification. It had ceased to be a scientific theory. It indulged in prophecies. ('Replies' 1974: 984–985) By contrast, psychoanalysis never achieved the status of a science because it claimed to be able to explain any and all individual situations. It pretended to fit all possible kinds of human behaviour. A scientific prediction is a *quantitative* forecast of some precise future event—like Eddington's prediction of the 1999 solar eclipse. Its precision allows its easy refutation. A prophecy is a *qualitative* anticipation of some major future event—like Marx's lifelong expectation of the outbreak of a socialist revolution. Its failure can always be explained away by some ad hoc manoeuvre. Marx's version of historicism is 'dangerous' because Marx presented the history of human societies as a deterministic affair. Like Comte before him, Marx read into the evolution of human societies a law of stages: feudalism precedes capitalism, which itself will give way to socialism. The events in *Hörlgasse* had turned Popper away from such social determinism.

The job of the social sciences is not to pronounce prophecies. Their job is to analyse social actions and institutions. Popper dubs this approach 'situational logic', a procedure which he spelt out in more detail in *Poverty of Historicism*. (It is close to Weber's methodology, which is also based on methodological individualism.)

Popper scrutinizes Marx's philosophy. He is sympathetic but sceptical. He welcomes Marx's sociological approach: it replaces Mill's psychologism by sociology and Hegel's idealism by economic materialism.[6] Marx's materialism shows that it is more important to understand the material conditions of society than the mental states of individuals. For Marx, of course, the material conditions determine the mental conditions. His famous phrase: 'being determines consciousness' ('Das Sein bestimmt das Bewußtsein') is an expression of his holism. The slogan expresses, in paraphrase, that the social conditions, in which individuals are caught, determine the way they think and act. It therefore also captures the focus on the material conditions in society. Popper rightly points out that Marx overemphasizes the importance of the material conditions. There is an interaction between economic conditions and ideas. (*Open Society* II: Chap. 15, §3, p. 107) It is ironic that Marx, who turned Hegel on his head, should minimize the role of ideas at the expense of material conditions. His ideas were to furnish the 'proletariat' with the motivation to overthrow capitalism. Popper therefore considers Marx's economic materialism to be 'an oversimplication'. But his 'sociological analysis of the class situation in unrestrained capitalism is admirable'. (*Open Society* II, 1945/[5]1966: Chap. 16) 'Unrestrained capitalism' or how much scope a government should grant the 'free market' is a topic to which Popper would return throughout this life. These reflections also make an urgent plea for democracy, which protects citizens against the misuse of power. He criticizes

---

[6] Political philosophers may find it surprising that Popper does not analyse Marx's central notion of alienation, without which his condemnation of capitalism and plea for socialism make little sense.

Marx and Marxists for their failure to understand the 'significance of democracy', as 'the control of the controllers'. (*Open Society* II: Chap. 17, §6) For Marx the institutions of a capitalist society are fatally flawed. They will vanish when capitalism collapses. Marx believed in revolutionary change and the dictatorship of the proletariat. Popper pleaded for the piecemeal reform of institutions. He had another reason to be suspicious of wholesale reform: it leads to the problem of 'unplanned planning'. (*Open Society* II: Chap. 17, §6; *Poverty* 1957/^{2}1960: Chap. III) Plans for a global reform of society, as envisaged in Marxism, invariably run into unforeseen problems, which need to be addressed. These unforeseen obstacles then require new planning, which was not part of the original plan. They may throw the original plan into disarray. In order to circumvent this danger, Popper prefers the piecemeal reform of institutions. They must be robust enough to prevent rulers from causing too much harm. (*Open Society* II: Chap. 17, §7) He distinguishes a radical form of Marxism, based on class rule, from a moderate version, which is compatible with democracy and capitalism. (*Open Society* II: Chap. 19, §3) Marx seriously underestimated the power of capitalism to reform itself. He seems to have ignored that in his own lifetime England introduced labour laws (Factory Acts 1847) to prevent the worst effects of unrestrained capitalism, such as child labour. Germany followed at the very beginning of the twentieth century.

Popper was not blind to Marx's failures, which he blamed on his historicism. That is, Marx's prophecy that the increasing misery of the proletariat would push capitalism over the edge; in line with the law-like succession of historical periods. The failure of capitalism would hasten the socialist revolution. However, such sweeping historical predictions are problematic. They encounter the problem of unplanned planning. The contingency of human agency prevents firm predictions. In *Poverty* he added that long-term predictions of human affairs are not possible since societies are not governed by 'social laws'.

Popper put his faith in democratic reform. But he praised Marx for his 'protest against unrestrained capitalism'. Marx's condemnation of nineteenth-century free market capitalism reveals a critical attitude, which Popper identifies as Marx's rationalism. However, there is a difference between *critical* and *uncritical rationalism*. (*Open Society* II: 229–230) Uncritical rationalism indulges in an absolute and blind faith in reason and traditions. But it cannot justify itself. There are passages in Popper's work, which are reminiscent of overconfident rationalism. For instance, he recommended his own 'critical' rationalism as the 'completion of Kant's critical philosophy'. (*Open Society* II: 231; *C&R* 1963: Introduction, p. 27; *Auf der Suche*^{2}1987: I.3, p. 60; *In Search* 1992: I.3, p. 48) In his fight against relativism and subjectivism he stressed the rationality of solutions. He recognized the role of emotions but spoke of an irrational faith in reason. (*C&R* 1963: Chap. 18, p. 357) Yet, he did not really believe in the omnipotence of reason. (*Auf der Suche*^{2}1987: Part III, §15, p. 232; *In Search* 1992: Part III, §15, p. 205) His *Critical Rationalism* puts the emphasis on criticism rather than justification. (Cf. Miller 2005, 2012: 63–65) It shares the Enlightenment belief in the power of reason but is critical towards its own solutions. It is a trial-and-error method. It is based on the belief in the unity of reason, argument and experience, impartiality, the evolutionary growth of knowledge and gradual social reform. (*Open Society* II: Chap. 24)

4.3.1 Individualism Versus Holism

Economist Friedrich August von Hayek (1899–1992). Nobel Prize 1974. *Source* Wikimedia Commons

The Marx chapter is entitled 'Marx's Sociological Determinism'. (*Open Society* II: Chap. 13) It hails Marx's greatest achievement as the 'autonomy of the social sciences', again a rebuff of psychologism. The problems of society cannot be reduced to speculations about human nature. Popper rejected the psychologistic method in the social sciences. He never deviated from the anti-psychologism, which is clearly stated in *Logic* (1959/1980: Chap. I.7, p. 44; *Logik* 1935/⁵1973: I.7, p. 18; cf. *Open Society* II: Chap. 14). He drew a distinction between 'subjective experiences', which cannot justify statements, and 'objective logical relations'. Psychology, he declared in his 1961 contribution to the conference held by the German Sociological Association in Tübingen, cannot be regarded as 'the basis of the social sciences'. Sociology is autonomous.[7] The social sciences must adopt the principle of methodological individualism, which makes its appearance in *Open Society* (I: 100–102). Methodological individualism, in contrast to Marx's holism, is a 'bottom-up' approach in the social sciences. It holds that all societal phenomena must be explained in terms of the attitudes and behaviours of individuals. Social groups are composed of such individual agents. Their behaviour can be coordinated or uncoordinated. But human activities are governed by social norms and guided by social values. They are embedded

---

[7] Popper (1969: 119; 1976: 101–102). With this lecture Popper became embroiled in what has been labelled 'The Positivist Dispute in German Sociology', which will be discussed later. Note that in his contribution to the dispute with the Frankfurt School, Popper argued that a model of a social situation must incorporate the institutional situation the agent faces when acting. (Popper 1969: 120–122; 1976: 103–104)

in traditions and institutions. Social facts are not reducible to psychological facts, because social relations include institutional aspects.

How can methodological individualism include the institutional aspects of social life? Since individual social agents operate within a network of social relations, such an approach has been characterized as 'institutional individualism'. (Agassi 1975; cf. Toboso 2001; Weinert 2022) According to this view only individuals have aims and responsibilities but institutions (cultures, states and traditions) are of primary importance to the social sciences. They exercise constraints on individuals but individuals also have the capacity to change social conditions. It is for this reason that trends exist in the social world, but social 'laws' do not. Popper's friend, the economist von Hayek, delivered a staunch defence of individualism in his influential book *The Road to Serfdom* (1944). He identified collectivist planning with totalitarianism and individualism with a competitive free market society.

> It is this recognition of the individual as the ultimate judge of his ends, the belief that as far as possible his own views ought to govern his actions, that forms the essence of the individualist position. (Hayek 2014: 73)

Like Popper, von Hayek was not in favour of *laissez-faire* capitalism. He recognized the existence of social goals but granted the state less power of interference than Popper. (Hayek 2014: 79) Both rejected 'collectivism' and therefore Marx's holism. Popper objected that groups, classes and societies are theoretical constructs and not empirical phenomena. (*Open Society* II: Chap. 14; *Poverty* 1957/²1960: Chap. 23; *C&R* 1963: 341; cf. Simkin 1993: 138–139) He held that there are 'no masses, only individuals'. (*Auf der Suche*² 1987: Pt. III, §15, p. 252; *In Search* 1992: Pt. III, §15, p. 222) And 'social wholes in the sense of totalities cannot be the object of scientific inquiry'. (*Poverty* 1957/²1960: Chap. III, pp. 74, 76–83) Both von Hayek's and Popper's defence of individualism must be seen against the backdrop of their experience with totalitarianism.

Their approach stands in stark contrast to 'methodological holism', which is a 'top-down' approach. This approach teaches that the whole is greater than its parts. Hence social macro-phenomena cannot be explained by a summation over individual actions. The French sociologist, Emile Durkheim, adopted this approach in his explanation of suicide: it was a societal problem. According to Durkheim society imposes constraints on its members, in terms of norms, values and traditions. Marx's famous dictum that 'being determines consciousness' also reflects a 'holist' commitment. For Marx class membership 'determines' the way of experiencing the social world.

It should come as no surprise that Popper, like von Hayek, wished to replace 'collectivism by individualism'. (*C&R* 1963: Chap. 16, §§6, 7; *Poverty* 1957/²1960: IV, §32) But both confuse holism as a method in the social sciences with the political doctrine of collectivism. It is a consequence of their attack on historicism and Marx's holism. Nevertheless, there are prior practices, traditions and institutional structures within which individuals must operate. Popper did not neglect the institutional aspect of social life. In his later work he emphasized the importance of structures in the understanding of the natural world. Some of his followers applied the notion of structure to the social world. (Shearmur 1996: 44, 125) Individual agents are necessary but not sufficient for the operation of the whole society. There are social entities which at any time transcend individual members of society. Think of language: the individual acquires the language spoken in a particular country. It exists before and after the life of each individual. A language would not survive if no speaker practised it. But its survival is not dependent on any particular individual. Similar remarks apply to traditions and social customs. As mentioned, Agassi (1975) renamed this approach 'institutional individualism'.[8] Methodological holism goes further: it holds that only social facts can explain other social facts. Large-scale phenomena must be understood by reference to social macro-phenomena.

In the final chapter (*Open Society* II: Chap. 25) Popper reflects on the nature of history. He denies that it has meaning. There are only histories. He echoes Vico's view that history is a human construction and hence contingent. Marx, of course, believed that the history of society had an aim, namely to bring about the total emancipation of humanity from oppression (alienation under capitalism.) *Open Society* anticipates some of the central ideas that Popper introduces in *Poverty of Historicism*. It also functions, like *Open Society*, as part of his war effort. *Poverty* contains important reflections on the nature of the social sciences. It also arose out of Popper's life in New Zealand, a country I know from personal experience (1988–1994).

4.4   Life Down Under (1937–1945)

In his *Autobiography* (1974: 96) Popper confesses:

> I liked New Zealand very much, in spite of the hostility shown by the University to my work, and I was ready to stay there for good.

This hostility seemed to have disappeared when the Poppers visited New Zealand from the end of February to the middle of May 1973, on the initiative of Alan Musgrave [330.15]. It turned into a trip around the world. Before they arrived, they stopped over in Bali for a holiday. They returned 'via Tahiti, Mexico City (Pyramids) and Miami (Dolphins), to Copenhagen and Ascona (where I visited my sister)' (letter to Feigl 17/09/73 [294.06]). Otago had invited Popper to a William Evans Visiting Professorship, and he was to deliver the Broadhead Memorial Lecture during twelve days in Christchurch. On his return he called Bryan Magee to tell him that he had received a 'royal reception in New Zealand' (letter of May 21, 1973 [322.19]).[9] I had been at *Victoria* University in Wellington for a number of years, when I heard that

---

[8] It is by no means established that individualism is the only defensible position in the social sciences. Brian Epstein, of Tufts University in Boston, won the prestigious 2016 Lakatos Award (*LSE*) for his defence of methodological holism in the social sciences. (Epstein 2015) Even a liberal writer like Jeremy Shearmur (1996: Chap. 5) sees in the existence of structural constraints in societies limits to methodological individualism.

[9] Popper felt ill on arrival in Christchurch (Hennie to Musgrave 03/08/73). Musgrave told the Poppers (27/07/73) that, according to reports he had received, the Broadhead lecture had gone very well [553.13; 563.01]. In 2002 a centennial conference was held in Christchurch in Popper's honour.

Popper had died on September 17, 1994. I felt that Popper's life and work should be celebrated in some way, since he had spent seven years in Christchurch. Fortunately, I knew Peter Munz in Wellington. (Fig. 4.2) He was a historian at *Victoria* University with an interest in philosophy. He was very supportive. Together we managed to organize an informal memorial event at *Victoria* University. Munz had studied under Popper (at Canterbury College in Christchurch) and Wittgenstein (at Cambridge) and had become one of Popper's disciples and lifelong friends. 'When Popper saw that I was genuinely interested in philosophy he invited me to his house [on Cashmere Hill, Christchurch] Saturday morning'. (Munz 1994) He regarded Popper as his 'intellectual father', to whom he owed 'half of his life' [330.10]. Like the Poppers, his Jewish family had been forced to leave Germany and emigrate to New Zealand in 1940.

The hostility to which Popper refers had to do with the remit of his appointment and the afore-mentioned Ivan Sutherland. He was expected to teach and not 'waste' his time on research. Research in his own words was regarded as 'theft from the working time for which I was paid'. (*Autobiography* 1974: 95) It is no surprise that Popper felt unappreciated and unrewarded in New Zealand. (Miller 1997: 377) True, he had a tendency to feel undervalued but on this occasion he did not exaggerate his difficulties. A special issue of the *New Zealand Science Review* (1991) contains a number of personal reminiscences by New Zealand academics who testify to this anti-intellectual climate. One contributor, H. Offenberger, confirms that 'Popper had been told by the university authorities of the day that he was stealing time by doing research when he was paid to lecture to students'. When *Open Society* was finally published Popper did not want 'Canterbury College mentioned, since they did everything to prevent me from writing the book' [300.04].

The contributors to the special issue also testified to Popper's unbounded enthusiasm and energy. His teaching style and ideas had an impact both in Canterbury and beyond. In a 'Personal Recollection', Williams (1991: 75–76) recalls that in 1938, in his second year at university, he decided to take a module in philosophy. He remembers that Popper was the only person teaching philosophy, and 'the other members of that department taught psychology'.

> It was a stimulating experience. Popper's lecturing style demanded absolute attention. He would stop anyone he saw taking notes and at the end of each lecture would dictate his own brief notes. His command of English was good…..

> As well as intellectual stimulation, I was moved by the fact that sometimes his wife, Hennie, would be in his office, conducting correspondence which they had with a number of Germans and Austrians who were trying to escape from their countries.

> Popper's energy was amazing. Not content with carrying the full burden of all the philosophy teaching, and carrying out research on several major topics, he found time to participate in the discussions organized by student groups on scientific and philosophical questions.

> He had a flair for presenting complex issues with such lucidity that both complete beginners and the relatively well informed were enlightened. A course he gave to a local Workers' Educational Association class, of very mixed background, gave me an insight into the quantum theory that three years of formal physics tuition had failed to produce.

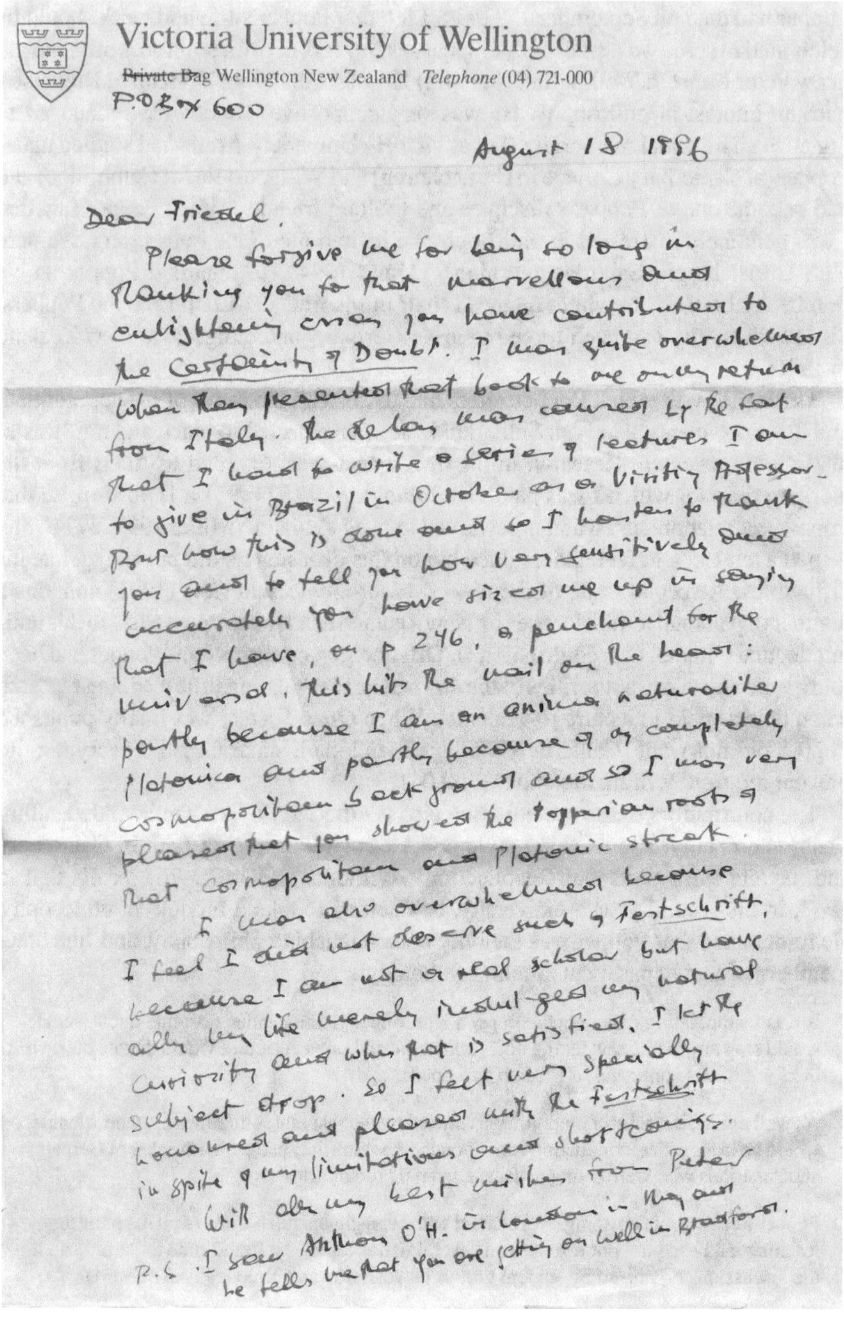

Victoria University of Wellington
Private Bag Wellington New Zealand *Telephone* (04) 721-000
P.O.Box 600

August 18 1986

Dear Friedel,

Please forgive me for being so long in thanking you for that marvellous and enlightening essay you have contributed to the Certainty of Doubt. I was quite overwhelmed when they presented that book to me on my return from Italy. The delay was caused by the fact that I had to write a series of lectures I am to give in Brazil in October as a Visiting Assessor. But now this is done and so I hasten to thank you and to tell you how very sensitively and accurately you have sized me up in saying that I have, on p 246, a penchant for the universal, this hits the nail on the head: partly because I am an animistic naturalite Platonica and partly because of my completely cosmopolitean background and so I was very pleased that you showed the Popperian roots of that cosmopolitean and Platonic streak.

I was also overwhelmed because I feel I did not deserve such a Festschrift, because I am not a real scholar but have always been like merely indulged my natural curiosity and when that is satisfied, I let the subject drop. So I feel very specially honoured and pleased with the festschrift in spite of any limitations and shortcomings.

With all my best wishes from Peter

P.S. I saw Anthony O'H. in London in May and he tells me that you are getting on well in Bradford.

Fig. 4.2 A copy of a personal letter, from August 18, 1996, in which Peter Munz thanks me for the essay I contributed to a Tribute in his honour

Popper's captivating lecture style continued to draw large audiences at the *LSE*. John Watkins, for instance, who began his studies for a B.Sc. (Econ) at the *LSE* in 1946, attended his first Popper Seminar on November 4, 1948. He was riveted because Popper's lecturing style was in a class of its own [360.02].

> A fellow student at LSE had told me it was worth going to Dr. Popper's lectures 'to hear the great man thinking aloud'. I went, and I was riveted. He had no notes or other paraphernalia. Ideas seemed to flow from him. They were put forward, not as propositions for the audience to consider, but as hard-won truths; his combination of seriousness, lucidity, and conviction had an almost hypnotic persuasiveness. The seriousness was lightened by touches of humour and happy improvisations. (Watkins 1997b: 663)

The experience turned Watkins into a faithful disciple who eventually became Popper's successor at the *LSE*. Agassi characterized Popper as 'the odd man out in the English milieu'. (Agassi 2008: Chap. 10D). But already in New Zealand Popper felt at first like 'a fish out of water in this remote, unconcerned environment'. (Munz 1991: 69) Munz wrote an obituary in the Wellington *Evening Post* (September 22, 1994), entitled 'Great Thoughts and a liking for milkshakes'. It describes Popper's rejection of the inductive method and the emphasis on error elimination. It also recalls Popper's social conscience:

> Popper had a deep social conscience, believing the State's main role was to protect the weak and underprivileged by use of "piecemeal social engineering".
>
> A dedicated and immensely efficient teacher, he explained even the most arcane problems simply.
>
> He hated any pomposity and professed an austere, even stern puritan morality, so much so that he played down subjective feelings and, with the exception of music, dismissed art and literature.

Cover of my copy of a collection of essays in honour of Peter Munz (1996). I contributed a chapter comparing the thoughts of Peter Munz and Karl Popper

Munz exaggerated for Popper was fond of authors such as Jane Austen, Anthony Trollope and Hugh John Lofting. (Watkins 1997b: 681) Amusingly, Munz adds that his only indulgence was food, especially milkshakes, which however did not continue in his later life.

> So he would drag his students in Christchurch from milk bar to milk bar, always insisting that the shakes were not good enough and we would have to try another.[10]

Despite the obstacles he faced at Canterbury College, Popper was not discouraged. He was working on his defence of the open society and pursued more technical work on logic. The Poppers founded a Committee for Refugees. Its remit was to help others who, like Peter Munz, had to flee from the Nazis. His study of Plato also led Popper to brush up on his 'scanty Greek'. He felt dissatisfied with the translations that were available in New Zealand at that time. His interest in Greek philosophy and cosmology never waned.

In 1998 Routledge published a collection of essays on Parmenides and Xeno-phanes (*The World of Parmenides*), for which Popper had written a Preface (dated February 27, 1993). The volume contains an early essay on 'Plato and Geometry' (1950) and republishes his essay 'Back to the Presocratics', which had appeared in *Conjectures and Refutations* (1963: Chap. 5). He states that 'all science is cosmology'. (Popper *C&R* 1963: 136; cf. Vaas 2019) It deals with the problem of understanding the world. 'Our cosmos bears the imprint of our minds'. (Popper, *C&R* 1963: 181) The remaining essays date back to the 1970s and 1980s. They reveal a much deeper sense of realism than the common-sense realism, which he often confessed.

Whence Popper's problems? Anthropologist Ivan Sutherland was a specialist of Maori affairs and promoted the Maori cause for greater integration and recognition in New Zealand society. In 1937 he was appointed to the Chair of Philosophy at the same time as Popper's appointment to a lectureship. Although the two men had a good relationship at first it became strained quite quickly. In 1940 (June 17) Popper asked the College to launch an investigation regarding 'my loyalty to the British cause' and his 'competence to lecture on the present war-issues'. His request was prompted by certain remarks 'made by the Head of my Department' which 'seem to me open to the interpretation that I am not implicitly trusted'. Rumours were swirling that a refugee had thrown a party to celebrate 'Hitler's victory over Holland'. Sutherland asked Popper whether he had been at this party. Popper had not attended the party, and, anyway, he thought that the rumours were ridiculous. But he considered that Sutherland's remark 'came dangerously close to spreading a rumour against myself'. On the same day (June 14) Popper had a conversation with Dr. Broadhead regarding the British war effort. Sutherland construed his remarks as 'not fitting for me as a non-Britisher to make'. Therefore Popper wanted 'an investigation into my loyalty and competence'. Four years later a dispute erupted about payments for marking and examinations at Victoria University in Wellington. Sutherland declared that Popper was only 'an assistant examiner' and so he received less money than other philosophy professors. Again Popper requested Council to review the situation since Sutherland

---

[10] Hennie wrote to Gombrich (29/07/43) that she and Karl went to the sea 'to eat as many icecreams as we could' to celebrate the completion of *Open Society* [300.02].

had insinuated 'that I am not carrying out my duties' (09/11/44). However, he was reminded that 'my protests are typical of behaviour which no Britisher or New Zealander would ever adopt'. Shortly afterwards (11/12/44) the Rector of the College informed Popper that he would not forward his complaint to the College Registrar, until they had time to discuss the matter in person. In the same letter, he reported that Dr. Sutherland disliked the 'increase in lecture hours for Logic and Ethics from three to four per week, [and] the placing of this option in Philosophy I or II (…) two days a week, when it would clash with other Arts subjects' [366.05].

Popper complained to Gombrich (09/04/45) that his position

…means spending one's life in a state of being bullied and of constant humiliation. And this for a salary which is and remains below what has to be spent, even if one does not buy books or other "luxuries".

The Poppers felt disadvantaged because of the big salary differences between professors and lecturers.

While all other members of staff got at least four increases during the last nine years, there is one, and only one, who has not got any.

Colleagues who protested against Popper's treatment were told:

We know that he is too good for this place. This we cannot help; and nobody will hold him if he goes elsewhere [300.04].

He had a mortgage on his house, had taken out a health insurance for the sake of Hennie and had to pay for the paper he needed to write his manuscripts. (Miller 1997: 377–378; Watkins 1997b: 655) Popper undertook some additional teaching at the local Workers' Educational Association in order to supplement his income. He worked under difficult circumstances. No research grants were available, and the College offered him no secretarial support. Library resources were scarce. It is reported that in 1934 the College Library contained no more than 15,000 books and worked on a budget of £340. Yet Popper gained the collaboration of a young economics lecturer, Colin Simkin, who became a lifelong friend. Popper asked the 24-year-old Simkin for assistance with his English and advice on the social sciences. (Simkin 1993) Popper possessed little knowledge of the social sciences apart from what he had learnt from his discussions with Julius Kraft and Karl Polanyi.

Before Popper left New Zealand he, Eccles and some other New Zealand academics published a pamphlet 'Research and the University' (July 1945), in which they emphasized that research and teaching are the two pillars of academic life. (Fig. 4.3) Eccles held a position at the Dunedin Medical School. As indicated above, he was one of the scientists who abandoned their inductivist beliefs and converted to Popper's method of falsifiability (Eccles 1974). At around this time Eccles and Popper also began to develop an approach to the philosophy of mind, which brought back to life some form of Cartesian interactionism.

Popper had come to New Zealand with the manuscript of the paper he had read at von Hayek's seminar at the *LSE* (1936). Work on this material, which was interrupted by the outbreak of the Second World War, eventually turned into *The Poverty*

RESEARCH AND THE UNIVERSITY

A statement by a group of teachers in the University of New Zealand

WE AGREE with Flexner's statement that research and teaching in the University should be ' conceived as hovering on the borders of the unknown, conducted, even in the realm of the already ascertained, in the spirit of doubt and enquiry '.[1] We regard research and teaching not as separate functions of a University teacher, but as complementary parts of a single activity.

We do not accept the point of view that teaching is the main function of the University, but even from this point of view it should be clearly seen that, as the highest school in the community, the University has inescapable duties to the most talented members of each generation, that is, to those capable of making contributions to the development of knowledge. We believe that the University fails in this obligation if the teaching it provides is not imbued with the spirit of enquiry as it is embodied in the tradition of research. Teaching which is not linked with research is on a lower plane, and does not stimulate the best intellects of successive generations of students.

The two activities of the University, teaching and research, should be co-ordinated and combined; and this fusion can, and should, be made a very natural one. Most research workers have a strong desire to teach since, to a considerable extent, teaching is necessary for their research work. Ideas are developed and clarified by imparting them orally to others, and by subsequent discussion in tutorials or seminars. Further, the real research worker feels the urge to hand on the torch of which he is the bearer, thereby perpetuating the tradition of the search for knowledge. On the other hand, teaching on a high level is impossible without research. One reason, in Hill's words, is that ' There is no zeal for research there is no vitality in teaching '.[2] but a more fundamental reason is that, without research, teaching can never reach right to the frontiers of knowledge.

Our postulate that teaching and research should be combined does not imply that in individual cases members of the University should not specialize in one of the two activities. We believe, however, that in a University such specialization should never be complete.

1) A. Flexner: *Universities American English German.*
2nd ed. 1931, p. 242.
2) *Third Congress of the Universities of the Empire,* 1926,
p. 81.

If the ideal sketched above is to be realised in New Zealand it is essential that the research function of the University teacher should be fully recognised. The commonly held view that the University is primarily a teaching institution should be abandoned, and the University should be looked upon as an institution in which the spirit of free inquiry is preserved and cultivated.

There are certain basic requirements which must be fulfilled if the University is to play its proper role as a research institution. These are: (1) the University must be supplied with adequate finance; (2) the academic staff must be large enough to ensure individual members sufficient freedom from teaching to undertake serious research; it must be realised that research often demands continuity of effort, and that it may temporarily absorb the whole energy of the worker; (3) the provision of the necessary space, together with the essential apparatus, and of technical and clerical assistance; (4) the provision of adequate library facilities, and, in particular, of periodical literature, on a greatly increased scale; (5) a break with isolationist tendencies, that is, the recognition of the need of contact with colleagues, within and outside New Zealand, by attendance at conferences, congresses, etc.; by visits to other research centres, and by regular sabbatical leave; (6) the provision of the means of publication by the institution of a University Press, or by monetary assistance; and (7) the recognition by controlling bodies that research activity should receive due reward in such matters as status and promotion.

The principles outlined above provide the material framework within which a University devoted equally to teaching and research could grow. Beyond providing this framework, no organisation of research within the University should be attempted. University research should be free. It should be directed merely by the initiative of the individual worker, and by his enthusiasm for his chosen problem.

These are the basic requirements, but they are inadequate in themselves. In addition, the right spirit must be present. What we need is the establishment and encouragement of a research tradition.

It is only in very rare instances that one sees the development of a new centre of research which cannot be traced back to the direct personal influence of a man who was brought up in one of the great schools of research. In a comparatively new country where such research traditions are rare, there is only one way of establishing them: they must be imported. This may be done either by sending promising research workers abroad or by importing research scientists from overseas. An excellent example of the successful importation of a research tradition into New Zealand is the establishment of the Otago School of Geologists by Professor W.N. Benson.

It is necessary to realise that as these methods will be successful only if the essential conditions for research are already provided, and if the research worker can count on understanding and recognition of his mission. New Zealand has lost many of its most brilliant men because it has failed to provide both the material and the spiritual conditions for their work. For example, Rutherford might have returned to New Zealand, instead of going to Montreal in 1898, but he knew very well that there was no hope of continuing or developing his research here. New Zealand has lost opportunities, in recent years, of attracting and keeping established and distinguished workers from Europe; opportunities from which many other countries are at present benefiting greatly. The widespread defeatist view (which at present dominates appointment policy) that New Zealand cannot keep a good man, must be given up. We hold that New Zealand cannot afford to lose a good man.

In this connection we may quote the following statement by Polanyi: ' Modern science is a local tradition and is not easily transmitted from one place to another. Countries such as Australia, New Zealand, South Africa, Argentina, Brazil, Egypt, Mexico, have built great modern cities with spacious universities, but they have rarely succeeded in founding important schools of research. The total current scientific production of these countries before the war was still less than the single contribution of either Denmark or Sweden or Holland. Those who have visited the parts of the world where scientific life is just beginning know of the backbreaking struggle that the lack of scientific tradition imposes on the pioneers .... However rich the fund of local genius may be, such environment will fail to bring it to fruition '.[3]

In order to remedy the situation as it exists in New Zealand a complete change in attitude is required. It must be recognised that a specialist might achieve much greater educational result by teaching his speciality, than by spreading his teaching over what is traditionally considered the balanced content of his subject. The view that it is the task of the University to hand to the students a definite body of examinable knowledge must be discarded.

If this is true, a complete revision of the examination system is required, with the object of giving greater freedom to the specialist approach of the teachers in different centres. Moreover, the role played by examinations in the University is at present greatly overvalued. We believe that at present examinations of the type at present in use ' are an insufficient and inconclusive test of the attainment of a University education '.[4] The educational task of the university must be taken much more seriously than its role in grading students.

The attitude adopted by us may appear to some a radical one, or even, perhaps, Utopian. Yet what we demand is nothing but the belated realization of principles laid down in the Reichel-Tait Report of 1925. From this report we quote the following statements: (1) ' The proper interaction of teaching and research is of the very essence of the highest education '; (2) ' Teacher and student in a University should be engaged jointly in a voyage of discovery in search of truth '; and (3), quoted from the evidence submitted by H.G. Denham, ' A teacher of science who is himself untouched by the research spirit is ... incapable of fulfilling the higher ideals of his position '.[5]

We believe that the great influx of students now in progress endangers University standards, so that this is an appropriate moment to recall and endorse these findings.

3) *Manchester Memoirs,* Vol. lxxxv (1941-43), No. 2
4) *Report of the Royal Commission on University Education in New Zealand,* 1925, p.14.
5) *ibid.* pp. 75-76.

Christchurch,
July 10, 1945.

R.S. ALLAN   (Professor of Geology, Canterbury)
J.C. ECCLES   (Professor of Physiology, Otago)
H.G. FORDER   (Professor of Mathematics, Auckland)
J. PACKER   (Professor of Chemistry, Canterbury)
H.N. PARTON   (Senior Lecturer in Chemistry, Canterbury)
K.R. POPPER   (Senior Lecturer in Philosophy, Canterbury)

(Reprinted with the kind permission of the Caxton Press)

THE CAXTON PRESS

Fig. 4.3 A scan of the Pamphlet, signed by Popper, Eccles and others, in defence of research in NZ universities. Popper wrote the text of the Pamphlet, with the exception of the first paragraph [300.04]. It caused 'quite a stir' as Popper wrote to von Hayek on August 1, 1945 [305.13]. The scan comes from my own copy of the issue of the NZ Science Review 46 (1989)

of Historicism . As Simkin recounted, Popper felt that the article was too abstract and that he, Popper, would start work on a companion article, 'Marginal Notes on the History of Historicism'. It became *The Open Society*. Popper struggled to get *The Open Society* published. (Gombrich 1999; Miller 1997: 379–380; Watkins 1997b: 656–660) Publishers in England and America rejected it. For instance, Cambridge University Press turned it down because it was too long and 'disrespectful to Plato' [300.02]. Eventually two of his Austrian friends—Ernst Gombrich and Friedrich von Hayek—came to his rescue. Thanks to their combined efforts Routledge published it 1945. Popper offered Gombrich payment for all the work he had done on the manuscript (08/05/44) [300.03]. But Popper was still looking for an American publisher. He turned to Russell for help. Russell was full of praise for *The Open Society* and recommended it to his American publisher Simon and Schuster. He described the book as of 'first-class importance' and a 'vigorous and profound defence of democracy' [345.14]. Russell's intervention was of no avail. Popper had to wait until 1950 when Princeton University Press published *The Open Society* in America. (Grattan-Guinness 1992) Although Russell generally agreed with Popper's hostility to Plato and Hegel, he also expressed some reservations. In a separate letter, dated 30/08/1946, he objected that there was no evidence for Popper's favourable view of Socrates and that Plato's theory of justice was more complex than Popper had made it [345.14]. The publishing history of Popper's *Poverty of Historicism* did not fare much better. It had been rejected by the journal *Mind*, before it was published in several parts in *Economica* (1944–1945), a journal of which von Hayek was the acting editor. An updated version appeared as a book in 1957.

Given his training and his rejection of psychologism, Popper had a preference for the natural sciences: theoretical physics and its history were later followed by an interest in evolutionary biology. Still, he was passionate about the methodology of science, a natural consequence of which was a comparison of the natural with the social sciences. He treated both Plato and Marx as social scientists. His rejection of their 'historicism' was a clear indication that there existed differences between the natural and the social sciences. But there were also similarities. It is useful to remember that disciplines like economics, sociology and psychology call themselves *social sciences* (*sciences sociales* in French, *Sozialwissenschaften* in German). My own assessment is that Popper's views on the social sciences have led to misunderstandings of his true position. It is much more nuanced than Comte's crude view of sociology as 'social physics'. In fact Popper explicitly rejected scientism in the social sciences. (*Poverty* 1957/^{2}1960: 60; *Objective Knowledge* 1972: 186; *Auf der Suche*^{2}1987/*In Search* 1992: I.1, II.12) He was opposed to 'physics envy'. (*Myth* 1994: 75) He was well aware of the nuances. His views on the social sciences are, in my view, surprisingly similar to Max Weber's methodology of ideal types. He did not refer to Weber in this particular context. His views on the social sciences, as expressed in *Poverty* and later, provide the philosophical backbone to his rejection of historicism.

4.5 Popper's Views on the Social Sciences

In *Open Society* (I: 57, 172, 239 [fn4]) Popper had drawn a sharp distinction between 'natural' and 'normative' laws. Natural laws or the laws of nature—the movement of the planets, the laws of gravity and thermodynamics—govern natural processes irrespective of human awareness. By contrast normative laws—patterns of behaviour and customs—are 'man-made'. They change over time; they can be reversed or ignored. Popper rejected the idea that law-like regularities are discernible in the history of human affairs. He calls *The Poverty of Historicism* 'one of the stodgiest pieces (of his) writing'. (*Autobiography*, Vol. I: 90) It is not 'exciting', he wrote to Schilpp (09/05/63) and that instead he should read *Open Society* [334.02; my translation]. But it contains some solid arguments, underpinning the rejection of historicism, as well as 'holism'. Out of these reflections grows a picture of what unites and what separates the natural and the social sciences.

The argument against historicism centres on the difference between *laws* and *trends*. The term 'laws of nature' refers to the regularities, which govern natural events and processes. The laws of science capture these natural regularities in the language of mathematics.[11] They can be modified, as Popper realized, if they do not agree with the empirical observations. (*Open Society* I: 239 [fn4]) The laws of nature cannot be reversed by human intervention. The tides are caused by the attraction of the moon and the sun. They cannot be stopped; their regular pattern cannot be changed. ('Time and tide wait for no man'!) But *trends* are reversible because they express patterns of behaviour. If it is a trend, say, that sadly 3,000 people die each year in alcohol-related car accidents, the government can and does introduce measures to reverse it. Philosophically speaking, a trend depends on and generalizes initial conditions. A trend sums over what many individuals do. But genuine laws of nature are independent of initial conditions. Newton's law of gravity states a general dependence of the gravitational force, F_G, on two masses, M_1 and M_2, and the inversely squared distance between their centres, $(\frac{1}{r^2})$. It is only when one wishes to compute the gravitational attraction between two particular bodies, say, the earth and the moon, that initial data need to be 'plugged' into the equation: the masses of the earth and the moon as well as the distance between their centres.

Popper observed that all laws are hypotheses but not all hypotheses are laws. When a hypothesis expresses a trend, it captures patterns of behaviour of a collection of individuals. 'Historical hypotheses are not universal', he continued, but are singular statements about a 'number of individual events'. (*Poverty* 1957/[2] 1960: Chap. IV, p. 107) Whilst there are no 'laws of social development', there are trends. Trends allow exceptions, in a way which genuine universal laws prohibit. An exception to a general law would either falsify the law or restrict its application. Genuine laws apply universally to a system, unrestricted in space and time. General trends may be transnational or restricted to individual societies. Trends depend on human behaviour, and human behaviour changes over time. Regularities in social life are 'not invariable': 'Historical relativity makes the methods of physics inapplicable to social life.' (*Poverty* 1957/[2] 1960: Chap. I, p. 6)

If trends are compatible with exceptions, what is the difference between trends and statistical laws? If half of an ensemble of atomic elements disintegrates over a given period of time, does this fact not make it a trend? The point about statistical laws is that the percentages they express are themselves invariable. The half-life of a number of atoms changes from element to element, from a few minutes to millions of years. It is the half-life itself, which cannot be changed. It is invariable.[12] Such statistical laws are also known in quantum mechanics. In the famous double-slit experiment a beam of electrons is sent through an inhomogeneous magnetic field. In the simplest case, the field splits the beam into two parts: with a 50% chance that

---

[11] See Weinert (1995): Introduction. Kepler's third law states that the average distance of a planet from the sun cubed (A^3) is proportional to its orbital period squared (P^2): $A^3 \cong P^2$. This is Kepler's discovery but the planets orbited the sun according to this regularity long before Kepler formulated his law (1619). Kepler's formulation of his three planetary laws constituted a radical break from the Greek tradition of circular orbits, which Copernicus had been unable to overcome. (See Weinert 2009: Chap. I)

[12] Ernest Rutherford, the discoverer of the atomic nucleus (1911), observed disintegration of elements in laboratory experiments. Half-lives in the order of millions of years must be computed. (See Weinberg 1993: 191–196)

an electron appears in the lower or upper part of the beam. The split beam leaves two marks on a recording screen. When electrons (or photons) are sent through the magnetic field singly and separately, they will still reproduce the pattern from the beam experiment. (Weinert 2004: Part I)

It is thus statistical laws, which are invariant (under identical conditions). Patterns of behaviour—trends—do not display such invariant features. Statistical patterns change with circumstances. Changing statistics are already apparent in the field of medicine. Penicillin is not as effective against infections as it used to be because of the evolutionary adaptations of bacteria. In social life such changing statistics are even more striking: drug addiction, traffic accidents, marriage and divorce figures change with social circumstances. The initial conditions of such trends vary.

The social sciences deal with trends, not genuine laws. Their generalizations are not 'falsified' by exceptions, and they lack precise numerical predictability. Yet Popper claimed repeatedly that there are fundamental similarities between the natural and the social sciences. On the one hand there are generalizations, which are analogous to the laws or hypotheses of the natural sciences. (*Poverty* 1957/^{2}1960: Chap. III, p. 62) On the other hand, the social and the natural sciences share a 'unity of method'. (*Poverty* 1957/^{2}1960: Chap. IV, §29) This hypothetico-deductive method consists of hypotheses and 'laws' from which 'deductive causal explanations' follow. An event to be explained (the *explanandum*) is derived from initial conditions in conjunction with an appropriate 'law' (the *explanans*). Such language implies that the social sciences can rely on universal laws, from which singular predictions could be derived. Popper repeated in a later publication that there is 'no real difference between the natural and the social sciences'. (*Myth* 1994, Chap. 8: 155) His formulations have led to misunderstandings. The picture is more nuanced. Popper contended that it must never be assumed that we have discovered truly universal laws (in the social sciences) but that is also true of the natural sciences. (*Poverty* 1957/^{2}1960, Chap. III: 102) Universal laws are hypotheses. In fact, he saw both similarities and dissimilarities between the natural and the social sciences. But the similarities should not be misconstrued as 'scientism'. (*Poverty* 1957/^{2}1960: Chap. III, p. 60) Sociology is not social physics. Both the natural and the social sciences start with problems. (*Myth* 1994: 155) Even in the absence of genuine predictive universal laws, the social sciences share two fundamental features with the natural sciences: *explanation* and *prediction*.[13]

On the question of **explanation**: In his paper on the 'Logic of the Social Sciences' (1976: 101) Popper characterizes the job of the social sciences as the description of the social world with the help of explanatory theories. This description includes

---

[13] One of the difficulties of discussing the social sciences in general is their wide range: from purely descriptive disciplines, like anthropology, to highly mathematical disciplines, like economics. Sociology lies between these two extremes because it employs both quantitative and qualitative methods. Popper was mainly concerned with sociology although he said later that he was mostly interested in economics: 'In fact, the only theoretical social science which appealed to me was economics. But like many before me I was interested to compare the natural and the social sciences from the point of view of their methods, which was to some extent a continuation of work I had done in *The Poverty*.' (*Autobiography* 1974: 96)

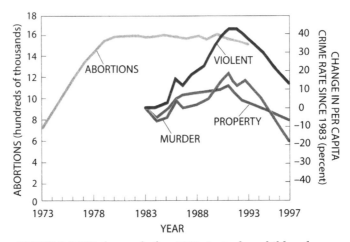

CRIME RATES *dropped after 1991, just when children born* after Roe v. Wade *would be reaching 18.*

Fig. 4.4 The Donohue-Levitt hypothesis. *Source* Scientific American (December 1999: 14)

an analysis of the 'problem of unplanned planning' and the complexity of human affairs. (*Poverty* 1957/² 1960: Chap. IV, §29, *Open Society* II: Chap. 14; *C&R* 1963: Chap. 4) Social scientists may not agree that their work is limited to an analysis of the unintended consequences of social action. Popper affirmed more generally that the social sciences must seek causal explanations. A famous example is the work of two US economists who linked the fall in crime rates in the US in the 1990s to the legalization of abortion in the 1970s. The legalization of abortion, they claimed, led to fewer unwanted births, unwanted children are more likely to commit crimes, so fewer unwanted children reduce the level of delinquency. (Donohue/Levitt 2001; Fig. 4.4)

On the question of **prediction**: The social sciences regularly make predictions. Economists predict the economic growth in particular countries, and even globally; sociologists predict that a baby boom will lead to a rise in juvenile delinquency. A self-fulfilling prophecy, for example, occurs when a prediction of the shortage of an item causes people to behave in a manner that makes the prediction come true (panic buying). A self-denying prophecy occurs when a prediction, like an impending epidemic, causes people to take precautions so that the prediction does not materialize. Such predictions are based on trends, which tend to be qualitative rather than quantitative. Popper held that the more trends approximate general laws the more trivial they become. (*Open Society* II: Chap. 25; *Poverty* 1957/² 1960: 150; *Alles Leben* 1996: Chap. 8) These generalizations become so obvious that they are not explicitly mentioned in social science accounts.[14] Less trivial patterns of behaviour (social or political trends) can form the basis of predictions.

Popper saw similarities between the natural and the social sciences because the natural sciences were more like the social sciences rather than vice versa. This claim refers to the hypothetical nature of knowledge. By the time he delivered his paper on the 'Logic of the Social Sciences' (1961) his hypothetico-deductive logic had morphed into rational criticism. The job of both the natural and the social sciences lay in the application of the trial-and-error method, the attempt to find solutions to problems. The social sciences employ a *situational logic* or *analysis*. (*Open Society* II: 97; *Poverty* 1957/²1960: 149) It is an analysis of social situations in terms of idealized models, for instance models of the typical consumer or conservative voter. Such models make rationality assumptions about human behaviour. It is assumed that social agents possess adequate information about their options and make rational choices in view of their aims. Empirical reality is then measured against the idealized model. Echoing Max Weber, Popper writes:

> By situational analysis I mean a certain kind of tentative or conjectural explanation of some human action which appeals to the situation in which the agent finds himself.

The aim is

> ...an idealized reconstruction of the *problem situation* in which the agent found himself, and to that extent make the action (…) 'rationally understandable', that is to say *adequate to the situation as he saw it*. This method of situational analysis may be described as an application of the *rationality principle*. (*Objective Knowledge* 1972: 179; italics in original)

Popper followed Max Weber's methodology of ideal types: models of *typical* individuals and situations.[15] Weber also embraced methodological individualism. In his monumental *Economy and Society* (1972: Chap. I.1) he characterizes action as social 'insofar as its subjective meaning takes account of the behaviour of others and is thereby oriented in its course'.

---

[14] In *Open Society* (Vol. II, 264–265) Popper mentions some examples. For instance, 'if we explain Caesar's decision to cross the Rubicon by his ambition and energy, say, then we are using some very trivial psychological generalizations which would hardly ever arouse the attention of a psychologist.'

[15] The term 'typical' refers to the modelling of behaviour in the construction of models in the social sciences. A situational analysis of social events requires the construction of a model, which captures what is typical of that situation. These social science models range from the 'typical consumer', the 'typical voter' to the 'typical democratic system' and the 'typical free market'.

Max Weber (1864–1920), influential German sociologist. *Source* Wikimedia Commons

The use of models is also vital in the natural sciences. Models allow scientists to describe concrete systems, to which abstract theories apply. There are models of the atom, the solar system or an ideal gas. Vital also is the awareness of problem situations. In a lecture at Oxford (November 3, 1967), Popper suggested that problem situations should be approached in the spirit of situational logic. Both the social and the natural sciences start

> …from myths – from traditional prejudices, beset with error – and from these we proceed by criticism: by the critical elimination of errors.

He proceeded to propose his familiar tetradic schema, where *TT* stands for tentative theory and *CD* stands for critical discussion. (*Myth* 1994: Chap. 7, §§1, 3)

$$P_1 \rightarrow \; TT \; \rightarrow \; CD \; \rightarrow \; P_2$$

Humans try to solve problems but they vary from discipline to discipline, and this is where the differences show up. To solve problems such as the unintended consequences of social action, the social sciences cannot rely on the availability of universal laws. There are only trends in social life and its history. The fundamental problem which the social sciences face is to explain and understand social events in 'terms of human actions and social situations'. ('Logic of the Social Sciences' 1976: 102; *Myth* 1994: 166) Human actions are guided by norms and values; they change over time. Methodologists therefore face a contrast between *universality* in the natural sciences and *particularism* in the social sciences. (*Poverty* 1957/²1960: Chap. III–IV; *Open Society* II: Chap. 25) The social sciences, which include history, deal with the scientific explanation of specific events. They are generalizing sciences but they do not give rise to powerful predictions, since their 'universal laws' are trivial. (*Alles Leben* 1996: Chap. 8) In fact, the description of 'concrete historical situations is like the statement of initial conditions in the natural sciences'. (*Myth* 1994: 166) As one commentator put it: 'Of all the sciences only general physics (…) seems to be exclusively concerned with the frame of nature'. (Kneale 1974: 212)

Popper placed biology in closer proximity to the social sciences than to physics. It too was concerned with the particular, namely the uniqueness of life.

The historical sciences must study or reconstruct not just human actions but also social situations. They must apply the method of situational analysis.

> The main point here was an attempt to generalize the method of economic theory (marginal utility theory) so as to become applicable to the other theoretical social sciences. In my later formulations, this method consists of constructing a model of the social situation, including especially the institutional situation, in which the agent is acting, in such a manner as to explain the rationality (the zero-character) of his action. Such models, then, are the testable hypotheses of the social sciences; and those models that are 'singular', more especially, are the (in principle testable) singular hypotheses of history. (*Autobiography* 1974: 93–94; italics in original; *Poverty* 1957/^{2}1960: §§31–32; *Open Society* II: 97)

Such statements reveal how close Popper was to Weber's methodology of ideal types.[16] According to Weber an ideal type is a 'conceptual construct which is neither historical reality nor even the "true" reality'. Rather it is a model, a construct, which captures the essential or typical features of historical epochs and social actions or events. (Weber 1949: 93) Ideal types are conceptual models, which allow the social scientist to delineate the 'salient features' of an economic, a historical, political or social situation. Social reality is measured against the model. In a paper on models Popper affirms that

> …models are even more important [in the social sciences]… They operate almost always by the method of constructing *typical* situations or conditions. (….) (T)he idea of a *social situation* is a fundamental category of the methodology of the social sciences. (*Myth* 1994, Chap. 8, §2, pp. 165–166; italics in original)

Real societal situations are then measured in terms of their deviation from the model situation. Economic models, for instance, attribute to economic agents' perfect rationality: 'typical' agents possess full knowledge of market conditions, of the available choices, of their own financial resources. They are capable of choosing according to a means-end rationality. The best economic decision—the best purchase or investment—is the one which fits their financial situation. Real people may approximate this model, although deviations are common. In *Poverty* and later Popper called this attribution of 'complete rationality' the 'zero method'. (*Poverty* 1957/^{2}1960: Chap. IV, 141–142; *Myth* 1994: 169, 172) The 'zero principle' generalizes to a 'situational analysis' or an 'analysis of the problem situation'.

The differences between the natural and the social sciences revolve around the problems and the methods of how to solve them. But if there exists a similarity between the natural and the social sciences, it consists in the demand that models in both areas must be testable. The models in the social sciences have, according to Popper, a 'low degree of testability'. (*Myth* 1994: Chap. 8. 4, p. 170) The reasons for this lower degree of testability are manifold: there are competing schools or

---

[16] This similarity between Popper and Weber has been noticed by other writers: Simkin (1993: Chap. 17); Shearmur (1996: 63–64); Hacohen (2000: 471–476). Popper's friend Ernst Gombrich (1974) would later apply situational logic to art and fashion. I suggested earlier that Plato's Ideal State could be interpreted in terms of Weber's ideal types.

paradigms in the social sciences (think of Adlerian, Freudian, Jungian psychology or Marxian and functional sociology); different interpretations are applied to the same data; social scientists have far less control over the test conditions than physicists have in laboratory experiments; and human agency is not reliably predictable. For instance in the famous Milgram experiments (1963), which tested the willingness of participants to obey authority figures, people's backgrounds (their education, their political views, their social status) could not be controlled. Social science results have hardly any claim to universality and lack the 'invariance' that is typical of findings in the physical sciences. Due to all these factors one may disagree with Popper's claim that the social sciences are 'less complicated than the physical sciences'. (*Poverty* 1957/²1960: Chap. IV, p. 140)

The question of what constitutes a problem situation came to full fruition in the 1960s when Popper faced Kuhn's influential thesis on the role of paradigms in science. Kuhn raised the issue of how rational the history of science really was. Popper insisted on the objectivity of problems and their solutions, according to the available tools at the time. (*Objective Knowledge* 1972: Chap. 4) This interaction between the perception of a problem, the availability of techniques to solve it and the proposed solutions expresses Popper's situational analysis. The logic of the problem situation thus came to encompass the history of science. It was no longer confined to the social sciences.

4.6 Arrival in the UK

When Popper arrived in the UK in early January 1946, he had developed clear positions both with respect to the natural and social sciences, as well as political philosophy. The next 20 years saw an extraordinary productivity and his rise to fame.

It was von Hayek who secured Popper a Readership at the *London School of Economics*. Hayek's support for Popper's candidacy was not really surprising, given the similarity between Popper's defence of the open society and von Hayek's qualified support of a free market. Von Hayek argued that central planning was incompatible with democracy. Both von Hayek's *Road to Serfdom* and the *Open Society* were political books and war efforts. Popper acknowledged many similarities between the two books but declared that his own book was 'infinitely inferior' to von Hayek's [305.13]. He felt a lifelong gratitude towards von Hayek. Long after his retirement from the *LSE* he called von Hayek 'a person of respect', a kind of 'father figure' [305.17]. The job offer itself reached Popper in the most unusual way. He recounted the story himself on the occasion of the first public lecture given to the Alumni of the *LSE* on June 9th, 1988:

> In 1944 I was travelling with my wife in a bitterly cold bus, returning from a skiing holiday in Mount Cook. The bus stopped in the middle of nowhere, at a snowed-in rural New Zealand post office. To my surprise, I heard my name called, and someone handed me a telegram – the telegram that changed our lives. It was signed F. A Hayek, and it offered me a Readership

at the London School of Economics. The appointment followed in 1945, and in 1949 I was given the title of Professor of Logic and Scientific Method. (*Propensities* 1990: 29; note that in his *Autobiography* 1974: 96 he correctly remembered that the job offer came in early 1945.)

The appointment was not straightforward. Already in July 1943 von Hayek asked Gombrich to provide him with a copy of *The Open Society*. His reason was that he wished to get Popper to the *LSE* [305.13]. On November 5, 1943, von Hayek approached Popper directly: did he wish to be considered for a Readership at the *LSE*? [305.13] Popper was enthusiastic about the prospect but feared that he would not get the job, due to his 'disappointing' publication list. Gombrich prepared the application on Popper's behalf. Woodger sent a letter of recommendation, so did C. M. G. Hight, the former Rector of Canterbury University College (07/12/43). Hight underlined Popper's 'remarkable clarity of exposition' and his excellent relations with his students and his colleagues:

> On his arrival he at once established, and has since maintained, most cordial relations with the rest of the staff of the College who highly appreciated his stimulating influence in faculty and committee meetings [406.03].

Popper had a different recollection so he also applied for other jobs: Balliol College (Oxford), Dunedin and Perth. In early 1945 he was offered a position at Sydney University, which he turned down. There was some opposition to the appointment of an 'enemy alien' but also because he was waiting for a decision on the *LSE* Readership. At the beginning of May 1945 the *LSE* had not yet made a final decision. The telegram, which reached Popper in the small village of Fairlie in the South Island of New Zealand, arrived on 28/05/1945 (not 1944) and read:

> UNIVERSITY LONDON OFFERS YOU READERSHIP LOGIC SALARY SEVEN HUNDRED AND FIFTY POUNDS LETTER POSTED PRINCIPAL UNIVERSITY LONDON [305.13].

The appointment as Reader in Logic and Scientific Method was initially for five years only. Popper worried that the move to the UK would cost him NZ £500 (=£400), which he did not have [300.04; 305.13]. Although he liked the tranquillity of New Zealand, the offer came as a great relief. The stress of writing the two books, the heavy teaching load and the dispute with Sutherland had taken a toll on his health. As Simkin (1993: 188) recalled,

> …(h)e was doing with very little sleep as he had spent most of his night getting the last part of *Poverty* and some of *The Open Society* into final shape. His blood pressure became very low, so that his doctor put him on to a variety of tablets and injections. On medical advice he also took two short holidays in the Southern Alps towards the end of 1944, and he felt better for them.

But the relief was temporary as he was soon suffering from other ailments. Yet he accepted Eccles's invitation, from April 1945, to give lectures on scientific methodology at the Medical School at Otago University, Dunedin, on the southern edge of New Zealand's South Island.[17]

---

[17] In the autumn of 1948 Eccles tried to entice Popper to accept the Chair of Philosophy at Otago, which John Findlay was vacating to return to South Africa. It was the prospect of a personal

Before they could leave New Zealand, the Poppers had to overcome administrative hurdles, such as exit permits from New Zealand. In late November 1945, they left from Auckland by boat, as they had arrived. They had sold their house but had only been granted 12-month visitor permits from the British authorities. Popper was worried and upheld his application to Dunedin. Simkin saw them just before their departure because he had come to Auckland to be interviewed for the Chair of Economics. He was appointed, he thought, largely because of Popper's intervention. Simkin recalls that

> Karl was taking, for reading on the voyage, *The Theory of Games and Economic Behaviour* which had just appeared from the pens of von Neumann and Morgenstern. Such was his idea of relaxation. (Simkin 1993: 188)

They reached England in early January 1946. They did not want to live in central London and considered Reading as a place of residence. Aware of the difficulties to find accommodation, Woodger offered Popper a spare room in his house (28/08/45). Popper accepted 'if things are desperate' but not wanting to outstay their welcome considered 'getting a caravan' (25/09/45) [363.17]. Although Popper was 'very thrilled' about his appointment, as he told Woodger (02/07/45), he was 'a little sad that I shall now, professionally, have to go on with the methodology of the social sciences: I am much more attracted to that of the natural sciences' [363.17]. Woodger, who signed some of his letters with 'Socrates', wasted no time in inviting him to speak to his local Labour Party Association, of which he was chairman (04/02/46) and the Fabian Society of Epsom (27/04/49) [363.17].

After arriving in England, Popper's first impression was one of shock at the unhealthy state of British philosophy. It was at that time under the spell of Wittgenstein's ordinary language philosophy. The animosity between Popper and Wittgenstein was born out of Popper's low opinion of linguistic philosophy. It led to a much-hyped confrontation between the two philosophers at Cambridge in 1946.

---

chair at the *LSE*, which persuaded Popper to decline the offer. Although he found life in London 'intolerable', as he wrote to Eccles [290.07], he preferred to stay at the *LSE*. He proposed John Passmore as the best candidate. Passmore occupied the chair from 1950 to 1955. Years later, when Otago approached Popper again for advice on a suitable candidate for the Chair of Philosophy, he recommended his former colleague, Alan Musgrave, who occupied the chair from 1970 to 2005. At 30 years old, Musgrave was the youngest professor at Otago at the time of his appointment.

Chapter 5
Popper's Life in England (1945–1960)

I am a rationalist of sorts. (K. Popper, *Conjectures and Refutations* 1963: 120)

The Poppers spent almost a lifetime in England, which according to Karl was 'a happy time'. (*Autobiography* 1974: 100) He was happy because he was working hard and was receiving recognition for his work. It was a productive period, in which he tackled old and new problems. He was developing arguments against determinism and the block universe (the view that the lapse of time is a human illusion); he replaced the traditional notion of confirmation by the notion of corroboration; his fight against 'subjectivism' led to a new interpretation of probability—the propensity interpretation; and whilst he had shunned the notion of truth in *Logik*, he developed Tarski's correspondence theory of truth into the verisimilitude account. His growing reputation meant that he was receiving many invitations to give lectures at foreign universities. And job offers came his way. In 1946 Viktor Kraft asked Popper whether he would consider returning to Vienna. On December 12, 1948, he again received a letter from Kraft, asking on behalf of the Philosophical Faculty of the University of Vienna whether he would be interested in taking over the Chair of Philosophy that had been occupied by Mortiz Schlick. Popper answered (26/12/48) that the offer was very tempting, 'if only the Nazi(s) were not there' [316.24; my translation]. He also had the prospect of a full professorship at the *LSE* with a salary of £1,500 p/a. Kraft, whose letters bear the stamps of the Austrian Censorship Office, answered (06/02/49) that the position in Vienna could not match this offer. So Popper stayed in England but their correspondence continued until Kraft's death in 1975.

F. Weinert, *Karl Popper*, Springer Biographies,
https://doi.org/10.1007/978-3-031-15424-9_5

Ludwig Wittgenstein (1889–1951). *Source* Wikimedia Commons

5.1 Wittgenstein's Poker

Just after Popper's arrival in England an incident occurred which caused a stir, at least in the philosophical community. Much has been made of this event, even leading to a book entitled *Wittgenstein's Poker*.[1] What happened? The Secretary of the Moral Sciences Club at Cambridge University had invited Popper to give a paper on the topic 'Are there Philosophical Problems?' At that time Ludwig Wittgenstein's ordinary language philosophy dominated the minds of philosophers at Cambridge and Oxford. Wittgenstein, a fellow Austrian, had famously declared that philosophy leaves everything as it is. This view stood in stark contrast to Popper's conviction that genuine philosophical problems existed and that philosophy could make a difference to the world. Popper accepted the invitation, being fully aware of Wittgenstein's position, namely that philosophical problems do not exist; there are only linguistic puzzles. Popper disliked linguistic philosophy. Already in *Grundprobleme* (1979: Chap. XI) he had stated that the preoccupation with language was a waste of time. And he had criticized Wittgenstein for his disregard of genuine philosophical problems. (*Grundprobleme* 1979: 416; 2009: 411) In a letter to his friend Viktor Kraft (15/06/70) he called Oxford's ordinary language philosophy the counterpart of Heidegger's philosophy: 'another form of the revolt against reason' [316.24; my translation]. Popper's insistence that there were philosophical problems worth pursuing—for instance, the problem of time—could not have marked a starker contrast with Wittgenstein's position.

---

[1] Edmonds, D./J. Eidinow, *Wittgenstein's Poker* (2001). In his conversations with Franz Kreuzer, Popper revealed that Wittgenstein did not threaten him with a poker, that he was only playing with it. (*Offene Gesellschaft* 1986: 36) Stephen Toulmin, in a letter to Bartley [272.09], confirmed that there was 'no suggestion of a threat.'

By accepting the invitation, Popper was therefore entering a lion's den. The lecture took place on October 26, 1946. According to Popper's own account of the incident Wittgenstein interrupted his talk repeatedly. When challenged to name some genuine philosophical problems, Wittgenstein dismissed all of Popper's examples out of hand. Popper, for instance, mentioned the problems of induction and of the validity of moral rules. Holding a fireplace poker in his hand, Wittgenstein asked him to state an example of a moral rule. Popper replied, 'not to threaten visiting lecturers with pokers'. This rejoinder infuriated Wittgenstein. He threw down the poker and 'stormed out of the room'. Reflecting on this incident in his *Autobiography*, Popper admitted that he 'went to Cambridge hoping to provoke Wittgenstein into defending the view that there were no genuine philosophical problems (…).' (*Autobiography* 1974: 98) Given their wildly divergent views, Agassi called Wittgenstein Popper's 'bête noire'. (2008: Chap. VI, p. 170) This may be an exaggeration. In *Grundprobleme* Popper discussed Wittgenstein's *Tractatus* (1921) as the intellectual inspiration of the *Vienna Circle*. There are frequent references to Wittgenstein in *Open Society* II. He told Wiesner (14/03/33) that 'according to his information' Wittgenstein had recently reacted 'catastrophically' to all criticism. 'He takes every criticism (…) as an assault on his physical existence' [Hansen 3.2; my translation]. In his later writings he continued to express his low opinion of ordinary language philosophy, which culminated in Wittgenstein's *Philosophical Investigations* (1958). After the Cambridge lecture Popper thanked Russell for the 'opportunity of co-operating with you (…) in the battle against Wittgenstein' (27/10/46). In his reply (18/11/46) Russell apologized to Popper for the poor manners shown by Wittgenstein and the Cambridge faculty [345.14].

5.2 Fallowfield

It seems that this encounter remained Popper's only brush with 'sensationalism'. Much more interesting is, however, that it reflects Popper's lifelong conviction that all activities start with problems, not only in philosophy and physics but in life generally. Disciplines are unimportant, he declared.

> We are not students of some subject matter, but students of problems. And problems may cut across the borders of any subject matter or discipline. (*C&R* 1963: 67; italics in original; *Objective Knowledge* 1972: 182 [fn30], 310; cf. *Realism/Aim* 1983/1985: Preface 1956, 159)

It is, in my view, one of Popper's abiding insights. And there were plenty of problems with which he was dealing. One was the complex problem of the relationship between entropy, irreversibility and time. He discussed it with Erwin Schrödinger, both in person, when he visited him in London in 1948 or 1949, and in their correspondence [347.18]. Schrödinger was an influential Austrian physicist, born in Vienna. He had won the Nobel Prize in physics (1933) and the Max Planck Medal (1937). He is famous for his formulation of the quantum-mechanical wave equation, which bears his name, and for Schrödinger's cat, his much-cited thought experiment involving a cat in a sealed box. Schrödinger had also made a foray into biology with the publication of his book *What Is Life?* (1944). The book investigates the question of the boundary between organic and inorganic life, between chemistry and biology. It contains an early theoretical description of the role of genetic material in inheritance, nine years before the discovery of the double-helix model of DNA structure. Popper disagreed with Schrödinger both on the question of the nature of time and the characteristics of life. These encounters and debates show that Popper's interests were wide-ranging, going far beyond the issue of the methodology of the natural and social sciences. He would remain focussed on these questions but after 1960 he made a decisive step towards evolutionary biology.

When Popper was considering applying for a chair in Cambridge, in 1948, Medawar told him frankly that certain people there were hurt by his tactlessness. (Watkins 1997a: 206, 1997b: 665) (But in 1971 Medawar became instrumental in Popper's election to a Fellowship at the Royal Society.) Still, several good things did happen to Popper in the first few years after his arrival in the UK. Thanks to the impact of *Open Society* and *Poverty* he was in great demand. In 1949 he was made 'Professor of Logic and Scientific Method' in the University of London. It was a personal chair. At the *LSE* he taught courses on Logic, Scientific Method and Problems of Philosophy. In the same year Popper received an invitation to give the William James Lectures at Harvard University. It was the first of frequent trips to the US. He revealed his earnings to N. R. Hanson[2] in a letter of 19/06/61: he was paid $6,000 for the eight William James lectures at Harvard; in 1956 he earned $2,000 for a fortnight's course of lectures at Emory University; and he received $9,450 from Berkley in 1962 [304.4].

On this first trip to the United States he met the Danish physicist Niels Bohr who was famous, amongst other contributions, for the first quantized model of the atom (1913). Bohr received the Nobel Prize for physics in 1922. In Princeton Popper also met his scientific hero Albert Einstein. As we have seen, Popper's admiration of Einstein's work dates back to 1920. Einstein left 'the greatest and most lasting impact' on him. (*Autobiography* 1974: 102) Popper was beginning to adopt indeterminism in physics as a general metaphysical attitude, a topic which he discussed with Einstein. He also tried to convince Einstein of a dynamic rather than a static view of time. Einstein had previously embraced the 'block universe' which he saw, like other physicists, as a consequence of the theory of relativity. (Weinert 2004) But at the time of Popper's visit Einstein was beginning to have doubts about the 'unreality of time'. Popper's arguments for the 'reality of time' were not falling on deaf ears. In his 'Autobiographical Notes' in the Schilpp volume, devoted to his work, Einstein (1949) clearly expressed a dynamical view about time. The view that the passage of time is real, and not an illusion, became a cornerstone of Popper's realism. As I will discuss, Popper was a 'realist', not just about time but also about the existence of the

---

[2] Norwood Russell Hanson was an American philosopher who invited Popper to come to Indiana University as a Visiting Professor for the spring semester of 1962–63. He would earn a stipend of $8,000 gross [304.04].

external world. But his realism became more sophisticated than the common-sense realism he often professed in his publications.

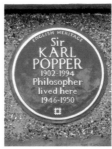

The Popper home in East Barnet, North London. *Source* Wikimedia Commons

Popper's home in Penn, Buckinghamshire. University of Klagenfurt/Karl Popper Library, Supplements 29; 2; 3. Copyright unknown

Popper earned enough money from the first US lecture tour to be able to buy a house outside London. In 1950 he and Hennie left their modest and noisy home in East Barnet, north London (16, Burlington Rise) to move to *Fallowfield*, a more comfortable and quieter place in Manor Close, Manor Road, Penn, Buckinghamshire, some 46 km north-west of London. Neither Karl nor Hennie had been fond of London life. In their new house the Poppers lived a quiet life, without even a radio [322.17]. But they had a cat called 'Chunky'.

David Miller (1997: 392) describes *Fallowfield* as 'the site of an intellectual cottage industry'.

> In his study, overlooking the lawn, Popper wrote always in longhand, producing complex coiling pages crammed with corrections and recorrections, cancellations and insertions. Hennie gave everything written her incisive critical attention, then typed it with four or five carbon copies, ready for the next round of revisions.

Bryan Magee called it a 'treasurehouse of unpublished manuscripts' and urged Popper to publish his work (19/08/69) [322.18]). There was a grand piano in the sitting room because Popper was fond of classical music, especially Bach. (He disliked modern music à la Schönberg). This love for music he had inherited from his mother. In a letter to Yourgrau's wife, Thella, Hennie revealed (18/08/76) that she and Karl made a 'pilgrimage once a year to Glyndebourne, the country house in Sussex where they play good operas during the summer months. We always go to a Mozart opera, and 3 out of 4 times it was Figaro' [364.12]. Books, however, he confessed were always more important than music. (*Auf der Suche* [2] 1987: 117; *In Search* 1992: 99)

The invitations were testimony to Popper's growing reputation. They occurred in his first decade at the *LSE*. Ayer, with whom he had corresponded since 1936, inquired (in 1954) whether he would be interested in applying for the Chair of History and Philosophy of Science at University College, London [270. 35]. It did not work out but he spent the academic year 1956–57 at the Center for Advanced Studies in the Behavioural Sciences at Stanford University. In order to obtain his visa, the American Embassy required 10–20 letters of support, testifying that he was not a communist. At the beginning of August 1956, Popper sent out requests for testimonials to prominent academics: M. Friedman, E. Gombrich, F. von Hayek, J. Kraft, J. H. Woodger and others [407.23]. On his return he became a Fellow of the British Academy in 1958, although he did not value this honour as much as his later Fellowship at the Royal Society (1976). In 1958–59, another attempt was made to entice Popper to return to Vienna. This time it came from Schrödinger (28/12/58; 29/01/59) who offered to use his contact with social democratic politician Bruno Kreisky, then Foreign Minister of Austria [347.18].

5.3 Discussions with Isaiah Berlin

At the end of the 1950s Popper got involved in a correspondence with Isaiah Berlin, whom Ayer had introduced to him in Oxford (on his second visit to England in 1936). Berlin was famous for his distinction between 'negative' and 'positive' freedom. This exchange of letters has not been much discussed.[3] But it throws an interesting light on Popper's indebtedness to Enlightenment ideas and ideals. Berlin, who occupied the Chichele Chair of Social and Political Philosophy at Oxford University, made this distinction in his inaugural lecture, entitled 'Two Concepts of Liberty'. He had sent a copy of the lecture to Popper, to which Popper replied on February 17, 1959. Popper expressed his agreement with Berlin's distinction but with some notable reservations. Berlin was in favour of negative freedom by which he meant non-interference of external agencies, like the state, in people's personal affairs. He associated negative freedom with liberalism, as defended for instance by the English philosopher John Stuart Mill. But Berlin warned against the dangers of 'positive freedom', which he associated with totalitarian thinking. He understood positive freedom as self-mastery, control over one's lower instincts. It does not at first seem to be particularly objectionable. What makes it disputable, according to Berlin, is the further implication that self-mastery should lead the individual to submit to a greater good, a higher aim and overriding common ideals. Such grander aims could be nationalism, communism or Rousseau's *volonté générale*.[4] In pursuit of 'positive freedom' the individual becomes subservient to the higher plan. Individuals sacrifice their own pursuits for the sake of a grander vision. Given Popper's own experience with the reaction of the Communist Party after the events in *Hörlgasse* in 1919, it is clear why he would share Berlin's concerns about the 'ideology of positive freedom'. But much worse for both Berlin and Popper was that the higher aim was imposed on individuals from above, by an organization (like the Party) of which they were a member. Berlin expressed fear that this would lead to despotism. Popper was of course as much opposed to totalitarianism as Berlin was to despotism. And Berlin's advocacy of negative liberty—the respect for the individual—appealed to Popper's enduring support for a philosophy of individualism and the welfare state. Popper was sympathetic to Berlin's praise of negative liberty (non-interference) and his warnings against the power of collective agencies. He himself held that

> …(t)he state is a necessary evil…. (*C&R* 1963: 350; cf. *Alles Leben* 1996: 227; *Lesson* 1998: 73)

---

[3] I would like to thank John Ackroyd for drawing my attention to this correspondence, which is discussed on the Persistent Enlightenment Project website (https://persistentenlightenment.com); Hacohen (2000: 510 [fn231]) mentions it briefly.

[4] Jean-Jacques Rousseau is the author of the famous book *Du Contrat Social* (1762), in which he affirms the sovereignty of the people.

Isaiah Berlin (1909–1997), historian of ideas. *Source* Wikimedia Commons

Despite their general agreement Popper suggested to Berlin an alternative inter-
pretation of positive freedom, which would be more in tune with 'negative freedom'.
What he had in mind was Kant's famous motto, with which he tried to capture
the spirit of the Enlightenment: *sapere aude.* Kant adopted it from the Roman poet
Horace. It can be translated as 'dare to know'. Kant meant it in the sense of 'having
the courage to use one's own reason'. This interpretation tallies well with Popper's
Critical Rationalism, which is thoroughly anti-authoritarian. Popper suggested to
Berlin that the 'search for truth—*sapere aude*—may be part of a positive idea of
self-liberation' [276.10]. Popper had admired Kant from a young age. He praised
him.

> …as a teacher of the Rights of Man, of equality before the law, of world citizenship, of peace
> on earth, and, perhaps most important, of emancipation through knowledge. (*C&R* 1963:
> 175)[5]
>
> Kant's 'own life was a story of emancipation through knowledge.' (*C&R* 1963: 177)

By adopting the motto *sapere aude* Kant had encouraged his readers to shake
off their self-imposed immaturity, their reliance on external guidance. To use one's
own reason and to critically examine claims which are peddled in the public sphere
requires training and knowledge. These aspects are captured in Popper's philosophy.
His Critical Rationalism and indebtedness to the Enlightenment led him to an anti-
authoritarian attitude towards science and other forms of authority.

> For I hold that science has no certainty, no rational reliability, no validity, no authority.
> (*Realism/Aim* 1983/1985: 222, cf. 259)
>
> So there are experts, but no authorities… (*Propensities*: 1990: 34; *Ich weiß* [2]1992: 24)

Individuals must not be pressed into the service of collective agencies in pursuit
of an imposed aim. Berlin was right to warn against the danger of totalitarian domi-
nation. The danger of totalitarianism looms larger when individuals are deprived
of their use of reason, when they are forbidden from contesting the decisions of

---

[5] These are themes which Popper defended again later as a public intellectual, which suggests to
me that he was a republican at heart.

the authorities. But a state that is concerned with the welfare of its citizens cannot forego interference into citizens' lives. Popper felt the need to defend the welfare state, which necessarily involves interference in the lives of individuals, be it through taxation, the criminal law or the social welfare system.

Finally, in 1959 his *Logik* appeared in English. After unsuccessfully looking for a translator, Popper asked Dr. Julius Freed and Lan Freed to translate the book. But, as he wrote in an undated letter to Bryan Magee he 'found the translation so bad that I translated it all over again, with their help' [322.16]. (It was not the first attempt, since Woodger had translated the book between 1946 and 1952 [363.17]). Popper added footnotes and appendices to the English translation but left the main text mainly unchanged. He also abandoned his original idea of adding a series of more substantial appendices to the *Logic of Scientific Discovery*. At Watkins's suggestion he decided to work the appendices into a separate book, written between 1951 and 1956. Watkins also suggested stylistic improvements. But it was not until 1982 that this work was published in three volumes as *Postscript*.

This was a busy decade for Popper, which saw him switch from determinism to indeterminism. In a letter to Feigl (10/12/73) he described indeterminism as incomplete determinism [536.10]. In its wake he developed, in conjunction with a deeper sense of realism, a dynamic view of time and his propensity interpretation of probability. 'I am *first* of all an indeterminist, *secondly* a realist, *thirdly* a rationalist'. (Popper, *C&R* 1963: 194; italics in original)

As always Popper was overworked. Watkins offered him a loan of £500 so that he could take a proper holiday in a 'sunny place' (17/03/54) [359.32].

5.4 Clocks and Clouds

When Popper arrived in Britain, he held in his arsenal a methodology of the natural and the social sciences. He was also a reformed socialist of a social democratic persuasion. His rejection of inductivism and historicism had shown him to be an intrepid and original philosopher. This made him a 'rare thinker' (Hochkeppel 1984) who cannot be associated with any particular school or ideology. (Parvin 2013: Chap. 3) He did not found a school of *Critical Rationalism*. It was a method, an approach to problems, which he now began to apply to a wide range of interconnected topics, for instance in his turn from determinism to indeterminism.

In *Logik* (§78) Popper had defended deterministic physics and its precise predictability. He now turned against determinism, not just in physics but as a general philosophy. A first indication occurred in *Open Society* (II: 85) where determinism was not considered a necessary part of the scientific method. He embraced indeterminism, which he describes as the position to adopt 'if at least one event escapes rational calculation'. (*Open Universe* 1982/1988: Chap. 1, §1; cf. *Quantum Theory/Schism* 1982/1992: 105) He moved from clocks to clouds. Adopting an indeterministic viewpoint has consequences for the rest of one's cognitive outlook. Indeterminism has to do with the 'asymmetry between future and past', as he wrote to

Schrödinger (15/01/53), [347.18]. He did not explicitly ask, like Newton or Leibniz, 'What is time?' He derided such 'what is?' questions as a form of Aristotelian essentialism, an inquiry into the essential nature of things. But his rejection of determinism meant that he embraced the 'reality of change' and the 'reality of time'. He also published a number of short papers on the arrow of time in the prestigious journal *Nature*, as we shall see. This cluster of ideas formed part of Popper's realism, that is the idea that there exists an external world, independently of human awareness of it. A further consequence of Popper's conversion to indeterminism was a new interpretation of probability: the propensity interpretation. What emerged then in this decade was a new 'worldview': the importance of dynamic change and the openness of the universe. Conceptually it follows on from his conjectural theory of knowledge and his disproof of 'iron laws' in history.

When in the 1960s, in response to challenges, Popper proposed a model of the growth of scientific knowledge, and delved into evolutionary theory and the mind–body problem, he did so on the basis of his belief in indeterminism and the reality of time.

5.5   From Determinism to Indeterminism

Popper accepted the traditional distinction between metaphysical (ontological) and predictive (scientific) determinism. (*Open Universe* 1982/1988: Chaps. I, II) These two versions of determinism can be found in Laplace's famous thought experiment involving a Demon (1814). The Demon surveys the whole universe—the cosmic landscape—in one fell swoop. It appears before his eyes like the frames of a film strip. All the frames already exist, giving a record of past, present and future events. The Demon simply moves his eyes, back and forth, along the film strip. Humans only possess knowledge of the past and present frames; future frames remain unknown. But they exist for the Demon. Furthermore, past events determine present events, and present events determine future events. There is an interlocking chain of events, linking past, present and future. For the Demon the history of the universe is laid out on a map of the cosmic landscape, irrespective of whether humans possess knowledge of it. The limited cognitive abilities of humans prevent them from seeing the landscape all at once, all the frames of the metaphorical film strip. The traditional image of a film strip or the newer image of the cosmic landscape captures what is meant by metaphysical or ontological determinism. All events already exist for the Demon. Popper conceived him as a superhuman scientist. It is important to include in this image the concatenation of events. Laplace's Demon insists on this causal chain. Past events are the cause of present events, which will causally determine future events. This dynamic aspect is better captured by the metaphor of the moving film strip; the image of the cosmic landscape emphasizes the eternal existence of events. No new events emerge, at least not for the Demon. Due to the cognitive limitations of humans, novelty seems to appear when new frames enter their visual field. They can also predict some events, at least into the near future. As the Demon can survey the whole landscape, there is no need for him to predict events. But human scientists project from knowledge of the past and law-like regularities to events which they expect to happen in the future. This aspect of predictability is the hallmark of scientific determinism. It cannot match, according to Popper, the perfection of a Demon. It is necessarily limited. (Weinert 2004; 2016)

Pierre-Simon Laplace (1749–1827). *Source* Wikimedia Commons

A philosopher may indulge in ontological determinism, a belief in the existence of a cosmic landscape. But as a scientist s/he would be a scientific indeterminist due to cognitive limitations. It is true of course that science can predict future events. The daily weather reports bear witness to this ability. Space missions, which land probes on distant comets, would be impossible without the ability to predict their trajectories. For instance, on November 12, 2014, the *Philae* lander, after having descended from its mothership, *Rosetta*, crash-landed onto comet 67P, after a ten-year journey from earth. It had travelled 3.7 billion miles. By contrast social actions are largely unpredictable because social scientists face Popper's problem of 'unplanned planning' due to the unpredictability of human agents. Meteorologists cannot confidently tell you what the weather will be like in Spain next year. Even cosmologists can only predict the future of the solar system to about 10 million years. The Demon, however, does not suffer from such limitations. He can see the dynamic unfolding of all events from the distant past to the far future. From the point of view of human scientists, he is a master of predictability.

Popper objected to both ontological and predictive determinism. He dismissed ontological determinism as a metaphysical doctrine, which cannot be falsified. He spent more time and energy on the refutation of scientific determinism. He hoped, thereby, to undermine metaphysical determinism. Popper became a metaphysical indeterminist, a conversion, which had started with his criticism of historicism. (*Alles Leben* 1996: 68; *Self/Brain* 1977: 32–4; cf. Dialogue X) Indeterminism, for him, was a cosmological fact. (*Quantum Theory/Schism* 1982/1992: §22, p. 181) The reader may express surprise. How could he embrace metaphysical indeterminism when he accepted that metaphysical determinism cannot be proved or disproved? At this stage in his career, however, Popper had come to understand that metaphysical positions can be criticized, even if not falsified. So he came to believe that indeterminism—the belief that past events do not lead inexorably to just one future event—was a more defensible position than determinism. It was more defensible because he had formulated three arguments against scientific determinism, which undermined metaphysical determinism.

1. One argument against scientific determinism is the lack of complete knowledge of initial conditions, from which precise predictions could be derived. This lack of precise knowledge of initial conditions could be due to two factors: (a) human cognitive limitations. Our physical theories only give us a simplified representation of the external world because they necessarily employ idealizations and approximations. Even if our theories imply a deterministic world, like Newton's mechanics, it could be a mistake to infer from them a deterministic universe. (Simkin 1993: §13) Or (b) it could be due to the existence of inherently unstable systems.

These two points can be illustrated by reference to some newer research into the chaotic behaviour of the solar system. (Laskar 1989) Popper himself referred to some older work, i.e. Hadamard's results of 1898, already discussed by Poincaré. (*Open Universe* 1982/1988: §14; Weinert 2016: 95)

Although the planets do not move erratically away from their periodic orbits, they do not follow the strict periodicity that Laplace had assigned to them. Planetary orbits move within a certain chaotic zone, of varying size, which is dependent on whether they are inner or outer planets. Their orbits within these zones are indeterministic but the zones themselves are stable. Determinism, however, claims that the trajectory of given mechanical systems can be exactly determined from a given current state both into the future and the past. For short time spans determinism seems to rule the planetary motions. For instance, calculations of Neptune's orbit show that Galileo must have perceived the planet 234 years before its discovery in 1846. But such apparent determinism of planetary motion does not last into the distant future. Indeterminism therefore characterizes the impossibility of determining the exact spatio-temporal trajectory of both classical and atomic systems. The French mathematician and physicist Henri Poincaré realized that their orbits may impose limitations on predictive determinism. Firstly, it may not be possible to specify the initial conditions accurately.

> If we knew exactly the laws of nature and the situation of the universe at the initial moment, we could predict exactly the situation of that same universe at a succeeding moment. But even if it were the case that the natural laws had no longer any secrets for us, we could still only know the initial situation *approximately*. If that enabled us to predict the succeeding situation with *the same approximation*, that is all we require, and we should say that the phenomenon had been predicted, that it is governed by laws.

Secondly, slight errors in the specification of the initial conditions will lead to breakdowns in predictions.

> But it is not always so; it may happen that small differences in the initial conditions produce very great ones in the final phenomena. A small error in the former will produce an enormous

error in the latter. Prediction becomes impossible, and we have the fortuitous phenomenon. (Poincaré 1908, quoted in Peterson 1993: 167; italics in original; cf. Penrose 1989: 220, 278)

The evolution of planetary systems can be predicted over 10 million but not over a hundred million years. Although the orbits seem to follow deterministic laws, they are not guaranteed to remain stable. Their stability and hence their predictability are not assured for eternity. In this sense, the motion of the solar system is 'chaotic' since it is not possible to compute and predict its evolution over a span of millions of years. (Laskar 1989) The motion of the planets constitutes an example where Laplacean determinism and perfect predictability do not coincide. (Cf. Penrose 1989: 226) Should this instability bother the Laplacean Demon? The Laplacean Demon possesses exact knowledge of the initial conditions and the laws of planetary motion. For the Demon the motion is perfectly predictable. But for his human counterparts, as Poincaré pointed out, the orbits are not perfectly computable. It would require complete and precise knowledge of initial conditions. If a small perturbation in the initial conditions will result in '100% discrepancy after 100 million years', this discrepancy constitutes an indeterminacy in the motion of the planets. If this discrepancy is not merely a limit in human predictability, which would not affect the Demon, but a genuine indeterminacy in the motions of the planets, it is to be expected that even the Demon would lose sight of the future trajectories of the planets. If the world is truly indeterministic, even on a large scale, the Laplacean Demon is powerless to restore deterministic behaviour. The Demon poses a challenge: he assumes that the universe is ontologically determined but his human counterparts fail to compute and predict the exact long-term evolution of the solar system. Is it truly indeterministic or does it only appear indeterministic to human eyes? It is possible, in theory, to assume metaphysical determinism and face predictive indeterminism in practice. Popper held that the world is truly indeterministic.

2. Popper's second argument refers to self-negating predictions. They affect classical physics. He interpreted the Laplacean Demon as a super-gifted scientist, surpassing a human scientist only by degrees of predictive ability. Popper then tried to show that the calculating precision, required to exercise Laplacean predictability, cannot be achieved in principle. To characterize predictive determinism, Popper introduced the idea of predictors. Predictors are predicting devices, which possess the ability to calculate the future and past states of the world from knowledge of present conditions and appropriate laws. Predictive determinism makes the calculating abilities of predicting machines (predictors) to acquire knowledge about the system's future evolution an essential feature of determinism. Insofar as the cognitive abilities of competent observers can be treated as approximations to the predictors, they also, like the predictors, should have the ability to predict all their future states. This ability requires the knowledge of boundary conditions and the differential equations, which govern the system. But predictors cannot calculate all their future states, because the information a predictor receives about its own state 'is liable to interfere strongly with

that state and thereby to destroy the predictive value of the information'. (Popper, 'Indeterminism' 1950: 189; *Open Universe* 1982/1988)[6]
3. Popper tried to show that the Special theory of relativity is not deterministic. It endorses the openness of the future. According to Einstein's theory, light propagates at a finite speed forming a past and a future light cone, at some event E. From events in past cones signals can be sent to events in future light cones. Events in past light cones can causally affect events in future light cones. Light and all other signals propagate from a source and disperse due to thermodynamic effects. It is practically impossible to reverse this dispersal.

> In physical terms, this asymmetry is established by the fact that from any place in the 'past', a physical causal chain (for example, a light signal) can reach any place in the 'future'; but from no place in the future can such an effect be exerted on any place in the past. (*Open Universe* 1982/1988: 58)

These arguments refer to the asymmetry of past and future.[7] They are central to both Popper's notion of realism and his views on the arrow of time.

Recall that Popper's arguments against scientific determinism do not really disprove metaphysical determinism, since by his own admission metaphysical positions cannot be decisively refuted. These arguments, however, make metaphysical determinism less plausible in his eyes. For if one sides with the Laplacean Demon, one has to assume an infinite chain of causal events which interlink the past, the present and the future. What we regard as random events, let us say the escape of a photon from an atom, is really determined by earlier events, which in turn are determined by previous causal events. So one ends up saying that an 'accidental' slipping on a banana skin can ultimately be traced back to earlier events, perhaps a dinosaur's sneeze or even the Big Bang. And what is regarded as free will is nothing but the interaction of neurons in our brains. It is perhaps no surprise that Popper's endorsement of indeterminism would eventually lead him to the mind–body problem!

In *Open Society* Popper had been contemplating broader concerns, which addressed his opposition to historicism and his belief in the openness of the future. He affirmed, quite correctly, that scientific discoveries and technological inventions cannot be predicted. If Newton had anticipated Einstein's Special theory of relativity, he would presumably have formulated it himself. Technological innovations can be prophesied—think of Leonardo's speculations about helicopters—but due to the lack of scientific knowledge they cannot be predicted. As science and technology have a strong impact on what is known at any one time, and their discoveries are unpredictable, future knowledge cannot be anticipated either. But this amounts to saying that change is unpredictable; even if change is planned, as Popper argued,

---

[6] Popper's argument is similar to his 'unplanned planning' argument against directed economies or societies: even if all future states are perfectly planned, contingencies will arise, which require new unexpected plans. Of course the new, modified plan will also run into problems of unplanned planning. These new 'unplanned plans' will interfere strongly with the previous plans and destroy their predictive value. (*Poverty* 1960: Chap. 21)

[7] For further discussions of Popper's arguments against determinism see Miller (1995); Watkins (1974); Weinert (2016: 11.2.2).

some of its consequences will be unpredictable. If this is correct, then change is fundamental. And if changing events are fundamental, then time is real, at least if the Leibnizian view is adopted, as Popper did implicitly, that the notion of time is related to physical changes in the universe. Such are Popper's reflections on determinism and indeterminism.

Of particular importance, from my point of view, is Popper's argument for the reality of time, which he developed in this period. It has not been sufficiently stressed in the Popper literature. It consists of two aspects: (a) the asymmetry of past and future and (b) the arrow of time. As I will try to show in Chap. 8, Popper's view of the reality of time is part of his defence of realism. And his defence of realism is part and parcel of his fight against subjectivism, especially in physics. He saw Heisenberg's famous uncertainty relations as an unwanted invasion of subjectivism into physics. His rejection of determinism and subjectivism led him to the propensity interpretation of probability. Determinism implies a subjective view of probability. How? If the world is predetermined, as Laplace's Demon asserted, but human cognitive frailty fails to grasp this ontological determination of the world, then any probabilistic statement expresses human uncertainties or lack of knowledge about the world. Humans do not possess the Demon's prescience. Popper, however, thought that probabilistic statements should be objective. They should not be a mere expression of our 'ignorance'. They should not be a reflection of our lack of knowledge. His propensity interpretation forms part of his realism about the physical world.

Looking back in 1974, Popper identified 'a few related problems—the propensity interpretation of probability, objectivity in quantum mechanics, and objectivity in the theory of time—which have between them taken up a considerable part of my life's work.' (Popper, 'Replies' 1974: 1117)

5.6 Probability

The best-known part of Popper's *Logic of Scientific Discovery* is his principle of falsifiability. Half of the book, however, is devoted to a discussion of theories of probability. In line with his general objectivism, Popper favoured objective over subjective interpretations of probability (the latter are in terms of degrees of belief). Probability plays a significant part in physics, especially in statistical mechanics and quantum mechanics. The roots of his opposition to a subjectivist understanding of quantum mechanics go back to 1934 (Fig. 5.1). At first, Popper aimed at a reconstruction of the frequency theory and then turned his attention to the role of probability in physics. In particular, he devoted a whole chapter to quantum mechanics. In the 1950s, he developed an alternative theory to the then current probability theories, which he dubbed the *propensity theory of probability*. Today the propensity theory has lost its connection with Popper's opposition to subjectivism. It has become an interpretation of probability in its own right.

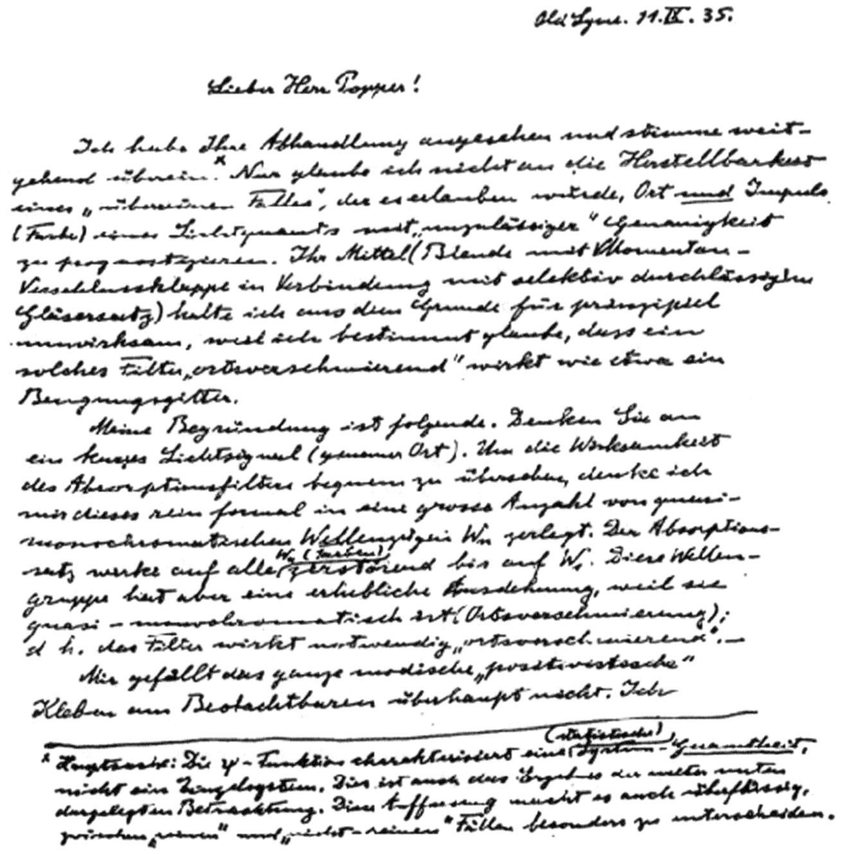

Old Lyme. 11.IX.35.

Lieber Herr Popper!

Ich habe Ihre Abhandlung angesehen und stimme weitgehend überein.* Nur glaube ich nicht an die Herstellbarkeit eines „überreinen Falles", der es erlauben würde, Ort und Impuls (Farbe) eines Lichtquants mit „unzulässiger" Genauigkeit zu prognostizieren. Ihr Mittel (Blende mit Momentanverschlussklappe in Verbindung mit selektiv durchlässigem Gläsersatz) halte ich aus dem Grunde für prinzipiell unwirksam, weil ich bestimmt glaube, dass ein solches Filter „ortsverschmierend" wirkt wie etwa ein Beugungsgitter.

Meine Begründung ist folgende. Denken Sie an ein kurzes Lichtsignal (genauer Ort). Um die Wirksamkeit des Absorptionsfilters bequem zu übersehen, denke ich mir dieses rein formal in eine grosse Anzahl von quasi-monochromatischen Wellenzügen W_n zerlegt. Der Absorptionssatz wirke auf alle W_n (farben) zerstörend bis auf W_1. Diese Wellengruppe hat aber eine erhebliche Ausdehnung, weil sie quasi-monochromatisch ist (Ortsverschmierung); d.h. das Filter wirkt notwendig „ortsverschmierend".

Mir gefällt das ganze modische „positivistische" Kleben am Beobachtbaren überhaupt nicht. Ich

* Hauptsache: Die ψ-Funktion charakterisiert eine ~~System~~ (statistische) Gesamtheit, nicht ein Einzelsystem. Dies ist auch das Ergebnis der weiter unten durchgeführten Betrachtung. Diese Auffassung macht es auch überflüssig, zwischen „reinen" und „nicht-reinen" Fällen besonders zu unterscheiden.

Fig. 5.1 First page of Einstein's letter to Popper, dated September 11, 1935, in which Einstein disproves Popper's thought experiment proposed in §77 of *Logik*. The copy appears in Appendix XII of modern editions of the book

5.6.1 Heisenberg's 'Uncertainty Relations' (1927)

Popper developed an objective interpretation of probability in response to what he saw as the invasion of 'subjectivism' into the interpretation of quantum mechanics (the physics of the atom). In *Logik* a general objectivism and realism prevail. In the company of a few physicists, such as Albert Einstein and Jean-Pierre Vigier[8], he opposed the then fashionable Copenhagen interpretation of quantum theory. It was defended by Niels Bohr, Max Born and Werner Heisenberg, all three winners of the physics Nobel Prize. Briefly it states that quantum systems do not possess

[8] He entertained a prolonged correspondence with French theoretical physicist Vigier about Einstein's realistic views on quantum mechanics and in opposition to the dominant Copenhagen interpretation. Vigier invited some physicists as well as Popper to contribute articles for *Le Monde* on these issues, which were to be published on January 16, 1983 [358.25; 566.17].

definite values before they are measured. The physicists of this school rejected the spatio-temporal description of particle paths. In particular, Popper targeted Heisenberg's 'uncertainty relations' as an example of such unwanted subjectivism. In *Grundprobleme* (1979: 199–200, 2009: 175) he had still accepted that quantum physics may offer 'good reasons' for the uncertainty relations. In a letter to Heisenberg, dated 26/11/34, Popper admitted that only a year earlier he would not have criticized the uncertainty relations [305.32]. It is notable that Popper often refers to Heisenberg's relations as 'uncertainty relations' rather than the more neutral term 'indeterminacy relations', which he used in his later work. The term 'uncertainty' refers to human knowledge. Heisenberg's 'uncertainty relations' state that it is impossible to obtain precise knowledge of the position and momentum of a quantum particle simultaneously. In fact, the product of both physical quantities is always greater than the Planck constant, h. In other words, if the position of the particle is known with precision, then knowledge of its momentum becomes imprecise and vice versa. This 'uncertainty' is expressed in Heisenberg's relation: $\Delta x \, \Delta p_x \geq \hbar/2$. (A similar relation holds for the pair energy/time.) It is still debated today whether the relation imposes limits on our knowledge or on the nature of the quantum world. At any rate, Heisenberg's principle can be used to explain the internal structure of atoms, i.e. the orbitals of electrons around the nucleus. Popper's realist instincts led him to a rejection of the Heisenberg relations as a limitation of what can be known about quantum systems. Popper did not accept that momentum would become imprecise when position was made precise and vice versa. His thesis was that the statements of quantum mechanics, including the uncertainty relations, were statistical statements [305.32]. He prepared an article for the journal *Naturwissenschaften* and sent Heisenberg the proofs. He also wrote several letters, between May 1934 and January 1935, to the physicist Victor F. Weisskopf [360.21]. He proposed experimental arrangements, which would support his thesis. Heisenberg discussed Popper's objections with his colleague Carl Friedrich von Weizsäcker, both then based at the university of Leipzig. Several letters were exchanged between November 1934 and March 1935 but the two physicists found Popper's objections unconvincing. They took Popper's considerations seriously but concluded, as Heisenberg wrote to Popper (19/12/34), 'that there is no experiment, which gets around the uncertainty relations in the way you want' [305.32; my translation]. Undeterred, Popper published a two-page synopsis in Volume **22**/48 of *Naturwissenschaften* (1934). The synopsis is spelt out in more detail in *Logic* (1959/1980: Part II, Chap. 9). (On December 6, 1934, Popper also sent a copy of his *Logik* to Heisenberg, which had then not yet officially been published.) Schrödinger applauded his assault on the Heisenberg relations (03/01/59) [348.18]. Popper even proposed a thought experiment, which would 'disprove' the Copenhagen interpretation. Its objective was to show that precise knowledge of both position and momentum could be obtained. Popper's thought experiment was much discussed by both philosophers and physicists, for instance by his friend Thomas D. Angelidis (2002). A postgraduate student at the *LSE*, George Zouros, wrote his PhD thesis on Popper's thought experiment and its implications. (Zouros 2007) Several years after Popper's death, two physicists found a way of turning Popper's thought experiment into a real laboratory experiment. (Kim/Shih 2007; cf. Qureshi 2013) Although the result was inconclusive, the discussions regarding Popper's objective views on quantum mechanics show that his contributions had some impact both in philosophy and physics. In the 1960s several physicists publicly endorsed his propensity interpretation of probability. (De Santos 2010)

Popper's doubts regarding Heisenberg's 'uncertainty relations' led him to a search for an objective interpretation. The orthodox view of Heisenberg's formulae, as part of the Copenhagen orthodoxy, is that they express restrictions on measurement. The physicist cannot measure both position and momentum simultaneously with arbitrary precision. As the precise trajectory of quantum particles cannot be determined, the Copenhagen view implied that the concept of particle paths was meaningless. Popper's objections are not directed at the mathematical formulation of the Heisenberg relations but at this standard interpretation as limits on what is measurable, real and can be known. Quantum theory is a statistical theory, so that the Heisenberg relations must be interpreted statistically. They are scatter relations, which means that they give rise to frequency predictions. Popper claimed that his interpretation was a consequence of the significance of Born's statistical interpretation of quantum mechanics.[9]

> It seemed obvious', he says in his *Autobiography* (1974: §18, p. 72; italics in original; cf. Quantum Theory/Schism 1982/1992: Introduction §3) 'that if quantum mechanics was to be interpreted statistically, then so must be Heisenberg's formulae: they had to be interpreted as *scatter relations*, that is, as stating the lower bounds of the statistical scatter....'.

Werner Heisenberg (1901–1976), famous quantum physicist who won the Nobel Prize in 1932. *Source* Wikimedia Commons

For Popper quantum systems are similar to the classical behaviour of particles which interact with each other, like the balls on a pin board. The particles follow

---

[9] Max Born received the Nobel Prize for physics (1954) for his so-called statistical interpretation of quantum mechanics. According to it the absolute square of Schrödinger's wave function can be taken as a probability density which gives the probability of finding a particle in a particular state (for instance its position or momentum). Born writes: 'Again an idea of Einstein's gave me the lead. He had tried to make the duality of particles—light quanta or photons—and waves comprehensible by interpreting the square of the optical wave amplitudes as probability density for the occurrence of photons. This concept could at once be carried over to the ψ-function: $|\psi|^2$ ought to represent the probability density for electrons (or other particles).' Born, 'Nobel Lecture' (1954: 262)

their trajectories, but the interactions disturb them stochastically. He never deviated from a 'realistic' view of electrons: they are just particles with position and momentum (whilst waves describe dispositional properties of particles). (*Quantum Theory/Schism* 1982/1992: 127) If that is so, Heisenberg's formulae do not refer to limits on measurements. They express limitations on our ability to prepare the states of a quantum system. The Heisenberg relations are statistical statements about a population of particles and the impossibility of preparing scatter-free quantum states. Their precise trajectories cannot be predicted, due to unavoidable scattering. From quantum theory 'we cannot obtain any singular predictions, but only frequency predictions'. This means that 'the statistical scatter makes it impossible to *predict* what the path of the particle will be after the measurement operation'. (*Logik* 1935/⁵1973: §76, p. 179; *Logic* 1959/1980: §76, p. 229; italics in original) Nevertheless, Popper insisted that it is possible, by a succession of measurements, to calculate the past history, i.e. the positions and momenta of particles.

> So it is only the precision of the *prediction*, which becomes blurred, or 'smeared' in consequences of the scatter relations, but never the precision of a *measurement*. (*Logik* 1935/⁵1973: §76, p. 180 [fn1]; *Logic* 1959/1980: §76, p. 231 [fn1])

By measurement he meant that trajectories could either be determined indirectly or calculated retrospectively.

> The scatter relations are frequency predictions about paths; and therefore these paths must be measurable – in precisely the same way as, say, throws of five must be empirically ascertainable – if we are to be able to test our frequency predictions about these paths, or about these throws. (*Logik* 1935/⁵1973: §76, p. 181; *Logic* 1959/1980: §76, p. 230)

Just as it is possible to determine the frequency of a particular throw of a die, so it should be possible to measure both energy and time, as well as momentum and position with greater precision than the Heisenberg relations allow. In fact, in order to test the scatter relations, '*we have to be able (...) to make measurements which are far more precise than the range or width of the scatter*'. (*Quantum Theory/Schism* 1982/1992: 54, 5; italics in original) Popper was inspired by classical analogies—the pin board and the tossing machine—to show how this could be achieved. From the scattering of a particle in a two-slit experiment and the trace on a photographic plate it would be possible to infer its momentum, given the deflection of the particle. (*Quantum Theory/Schism* 1982/1992: 144–47, 153)

After the development of his propensity interpretation, Popper came to see the Heisenberg relation as 'a singular probability statement', which reveals the propensity of a particle to scatter. Instead of referring to an 'aggregate of particles', he preferred to speak of an aggregate or sequence of 'repetitions of an experiment undertaken with *one* particle (or *one* system of particles)'. (*Logik* 1935/⁵1973: §75, p. 175 [fn*1]; *Logic* 1959/1980: §75, p. 225 [fn*1], italics in original; *Quantum Theory/Schism* 1982/1992: Chap. III, §16, p. 144) Instead of considering the 'number of cases', he focussed on the 'sum of the weights of the cases'. These 'weights of the possibilities (or of the possible cases)' were regarded as '*measures of the propensity, or tendency, of a possibility to realize itself upon repetition*'. (*Quantum Theory/Schism* 1982/1992: 70, italics in original; cf. 128–9; cf. Simkin 1993: § 10)

These characterizations marked a departure from Popper's original endorsement of Born's statistical interpretation, which did not assume that a particle had a hidden propensity to produce a probability. The statistical distribution, which Born had in mind, referred to '*actual* relative frequencies for a *finite* sequence of repetitions'. In his correspondence with Einstein, Born explained what he meant by the statistical interpretation of the wave function:

> To say that ψ describes the 'state' of one single system is just a figure of speech, just as one might say in everyday life: 'My life expectation (at 67) is 4.3 years.' This, too, is a statement about one single system, but does not make sense empirically. For what is really meant is, of course, that you take all individuals of 67 and count the percentage of those who live for a certain length of time. This has always been my own concept of how to interpret $|\psi|^2$. (Einstein/Born 1971: 186)

Popper's appeal to Born's statistical interpretation is puzzling for Born supported the Copenhagen interpretation of quantum mechanics, of which the Heisenberg relations are a central part. In his *Nobel Prize* lecture (1954) Born explicitly stated that the Heisenberg relations helped him to establish the statistical interpretation. A consequence is, he held, that classical determinism must be abandoned, as well as the 'naïve view of reality', according to which the position and momentum of quantum particles can be established simultaneously. At the end of his lecture Born concluded that he was in favour of retaining the particle concept but not in the classical sense of the word. Rather he appealed to the mathematical concept of invariants in transformations. 'Every object that we perceive appears in innumerable aspects. The concept of the object is the invariant of all these aspects'. (Born 1954: 266) But Popper continued to think of quantum particles in classical terms; then it was natural to assume that they had precise trajectories, even though their prediction became 'blurred'.

Max Born (1882–1970). *Source* Wikimedia Commons

A general difficulty with Popper's interpretation of quantum mechanics is that he treated the behaviour of atomic particles along classical lines. He recognized that there are dissimilarities between the classical and the quantum world. But he tended to emphasize the similarities. He saw an exact analogy (the pin board analogy) between the behaviour of ordinary balls meandering down a pin board and the famous collapse of the wave function. (*Quantum Theory/Schism* 1982/1992: 74) Quantum mechanics, however, constitutes a radical departure from Newtonian physics. It has many features which do not correspond to classical mechanics.[10] One of the difficulties is that it cannot explain characteristic properties of quantum systems, like entanglement, the superposition principle, Pauli's exclusion principle and the difference between bosons and fermions. In particular the new field of quantum information, which encompasses quantum cryptography, teleportation and quantum computation, relies on the quantum–mechanical properties of superposition and entanglement.

In his opposition to the 'subjectivism' of the Copenhagen interpretation, Popper treated quantum particles like classical particles. He reinterpreted the uncertainty relations as consequences of classical physics. His rejection of the Copenhagen view was a result of his objections against subjectivist interpretations of probability. In 1957 and 1959 he proposed his own objective interpretation of probability, which he dubbed the *propensity interpretation.* He promoted it both as an objective theory of probability and as a contribution to his ongoing fight for an objective interpretation of quantum physics. He claimed that the propensity interpretation could solve the paradoxes of quantum mechanics, such as the 'collapse of the wave function' and the question of 'quantum jumps'. He saw the elimination of 'irrational or subjective' elements as the main argument in favour of the propensity interpretation. (Popper, 'Propensity Interpretation' 1959; *Realism/Aim* 1983/1985: 351f; *Quantum Theory/Schism* 1982/1992: 68–74)

On a historical note it is worth remarking that, apart from Peirce, both Henry Margenau (1954) and Werner Heisenberg (1958) developed similar approaches to Popper's. (See Suárez 2007; 2020: §7) But Popper does not reveal whether he was aware of at least Margenau's work. Mauricio Suárez expressed his surprise in an email (01/06/2020) to me:

> Popper is not always very good in acknowledging his debts - other than to Einstein, Darwin, etc. I would suggest there are two kinds of influences on his work on propensities that are undeniable, however. One is Henry Margenau's 'latency school', which was very well known in the 1950's. Margenau was a prominent contributor to Philosophy of Science from its origins (he was founder and president of the association) and I find it impossible that Popper did not know of this work. The other, looming in the background is of course Peirce - which Popper does mention but not often in the context of propensities. I think the most coherent understanding of propensities is as theoretical non-observable properties (Margenau), postulated abductively as real properties that can explain statistical phenomena (Peirce). In his best moments, Popper comes very close to this view, as you point out.

So propensities are real physical properties arrived at by a process of abduction. Divorced from its origin, i.e. Popper's worries about the Heisenberg relations and the subjectivity of the Copenhagen interpretation, the propensity interpretation has established itself as an alternative interpretation of probability. (Miller 1994: Chap. 10, 2004; 2016; Suárez 2007; 2009; 2018; 2020)

---

[10] Popper denied the reality of quantum jumps. (*Quantum Theory/Schism* 1992: 137f) In *Realism and the Aim of Science* (1985: 360), he claims that the two-slit experiment can serve as a crucial experiment between the statistical and propensity interpretation.

5.6.2   The Propensity Interpretation of Probability

The propensity interpretation takes probability as a measure of physical dispositions or tendencies to realize what is physically possible. Popper chose this label because he wished to emphasize that it is possible to estimate the 'measures' or 'weights', which are attached to such dispositions or tendencies. (*Realism/Aim* 1983/1985: 358; cf. Gillies 1995; Miller 1994: Chap. 9; 1995) But these dispositions do not reside in things:

> They are not properties inherent in the die, or in the penny, but in something a little more abstract, even though physically real: they are relational properties of the total objective situation; hidden properties of a situation whose precise dependence on the situation we can only conjecture. (*Realism/Aim* 1983/1985: 359; *Quantum Theory/Schism* 1982/1992: 128-29)

The propensity interpretation, which partners with the classical frequency interpretation of probability, introduces into the discussion of probability the distinction between visible *appearance* and underlying *reality*, which Popper found in Kant and Parmenides. As I will discuss later, Popper came to hold a quasi-structural view of reality. In terms of probability this means that the finite, relative, measurable frequencies—the appearances—are the result of hidden physical dispositions or tendencies. Popper suggests that 'propensities' may exist—just like forces, or other abstract or 'occult' physical entities, introduced in order to explain the known by the unknown. (*Open Universe* 1982/1988: 105) It has not been sufficiently observed that propensities as hidden realities are part and parcel of Popper's general (quasi-structural) realism. Popper does not demote the appearances; appearances and underlying realities are (at least since Newton) equally real. Propensities are physically real, like dispositions, but they attach probability to a single event as 'representative of a *virtual* or *conceivable sequence* of events, rather than an element of an actual sequence'. (*Open Universe* 1982/1988: 287; *Quantum Theory/Schism* 1982/1992: 70–1; Miller 1997: 387–8) The propensities produce the observable frequencies, like Born's probability density. Popper compares propensities to Newtonian forces, i.e. as unobservable dispositional properties of the physical world. According to the propensity interpretation the expression $p\,(a,\,b) = r$ should be read as the assertion that the conditions, b, have a propensity, r, to produce state of affairs, a.[11] Popper saw it as an advantage of his interpretation that it could be applied to single cases or rare events. It allows single systems 'to have objective tendencies or propensities of varying strengths or weights or intensities (…).' (*Quantum Theory/Schism* 1982/1992: 129) There are single-case propensities, like the decay of a particular radium atom at a particular time or the passage of a single electron through a two-slit apparatus.[12] Thereby he

---

[11] *Realism/Aim* 1983/1985: 291. Alternatively, the statement means that the 'propensity of a state of affairs (or the conditions) b to produce a equals r', where r is a real number. (*Autobiography* 1974: 170 [fn178])

[12] Gillies (2000: Chap. 6; cf. Suárez 2007: 430–31; Suárez 2010: Chap. 7, 9.2) distinguishes long-run from single-case propensity theories. The former are associated with finite, repeatable conditions, which result in frequencies; the latter produce particular results on particular occasions, like, for

hoped to remove the 'subjective character' from quantum theory, as in the orthodox reading of Heisenberg's relations. Popper's objection to the Heisenberg uncertainty relations can be seen in the light of the distinction between observable phenomena and underlying reality. The particles possess both position and momentum, but these cannot be jointly observed because of the unavoidable scatter.

Propensities are relational properties of a total objective situation (like an experimental arrangement). Propensities are not frequencies; rather, they are 'dispositions to produce frequencies'. (Popper, *Realism/Aim* 1983/1985: 397; *Quantum Theory/Schism* 1982/1992: 71, 129–30; 'Replies' 1974: 1130) The underlying propensities are responsible for generating computable frequencies. Defenders of the propensity interpretation of probability often share this view:

> Propensities explain and ground single-case objective chances; chances are represented as probabilities in statistical models; and these models are, in turn, tested by frequency ratios in sequences of experimental outcomes. (Suárez 2018: 2; cf. Miller 2016)

Popper regarded the decay law and the double-slit experiment as the best arguments in favour of his propensity theory. (*Brain/Self* 1977: 26 [fn4]; *Quantum Theory/Schism* 1982/1992: §18) Atomic elements have a tendency to decay, and the rate of decay is determined by the Rutherford-Soddy formula (1902). The decay of a single, unstable neutron into a proton, an electron and an anti-neutrino happens in twelve minutes *on average*. The decay law only states the half-life of a population of particles and predicts, for each type of atom, when half of an initial ensemble will have decayed. Precise predictions of when individual particles decay are impossible. Only their average decay time can be computed. This average is expressed in the decay law. Popper noted that the decay law suggested an alternative interpretation of probability: 'each nucleus has a *tendency* or *propensity* to disintegrate, dependent upon its structure. This tendency or propensity can be measured by the *"half-life"*, a constant characteristic of the structure of the radioactive nucleus'. (*Self/Brain* 1977: 25; italics in original. Note the reference to *structure*, which became a central element of his more sophisticated notion of realism.) In Popper's interpretation, it is the experimental situation which reveals that, say, a single, unstable neutron has an underlying *propensity* to decay in 12 minutes on average, but the decay law determines the observable frequency of decay, i.e. the half-life of an ensemble.

Propensities differ from conditional probabilities—statements like $p(a/b)$. Conditional probabilities are symmetric: if there is a conditional probability $p(a/b)$, then there is a conditional probability $p(b/a)$. It can either be asked 'what is the probability of a smoker developing lung cancer' or 'what is the probability of a lung cancer patient being a smoker'? Propensities are time-asymmetric, since a disposition must be in place before it can be actualized. (Cf. Suárez 2020: §9.2) By contrast conditional probabilities may either be temporally asymmetric (heavy smoking precedes lung cancer) or a and b may occur simultaneously (heavy seas and the sinking of a ship). As dispositional properties, propensities may not be expressed in quantitative

---

instance, in coin tossing. Popper merged the two senses: a probability is a measure of a virtual frequency, which needs to be corroborated by the observation of actual frequencies.

laws, like Newtonian forces: $F = ma$. Popper tried to address some of these problems in his 'Replies to My Critics' (1974: 1125–39). If propensities do not obey the formal properties, to which conditional probabilities are subject, it is still possible to interpret them in a 'causal manner'. This seems to have been Popper's take on propensities, since he distinguished them from frequencies. He regarded his propensity interpretation as a generalization of the notion of causality. 'We can then causally explain (in a generalized and weaker sense of "explain") a is due to the presence of b, even if r does not equal 1'. (*Autobiography* 1974: 170 [fn178])

Propensities, according to this view, 'generate' the observable frequencies. But how? Even if the demand of formal definitions could be satisfied, there still is the problem of testability. In his replies to his critics Popper goes as far as musing that the propensity interpretation is a 'metaphysical' idea', as some critics had suspected. ('Replies' 1974: 1129; cf. Keuth 2005: §8.5) It is a coherent interpretation, yet, at the same time, Popper always maintained that the existence of propensities could be tested, at least sometimes. But he did not spell out how such tests could be carried out. The example of the decay of a sample of radioactive elements illustrates the problem. The only evidence for the propensity is the frequency of its decay. But in his earlier work Popper had dismissed such ad hoc reasoning as unacceptable. Reconsider the fictional example:

> The sea is rough today. Why?
>
> It is rough because Neptune is angry.
>
> How do you know that Neptune is angry?
>
> Don't you see that the sea is rough today?

In this example, as Popper pointed out, the only evidence for the explanans ('Neptune is angry') is the explanandum itself ('The sea is rough today'). What is required, however, is independent evidence for the explanans, i.e. Neptune's anger. In the same way statistical outcomes should not be used as evidence for the underlying, 'hidden' propensities. Popper refers to the law of large numbers as a 'bridge from propensities to statistics'. (*Logic* 1959/1980: §48 [fn6]; cf. Miller 2016: §2.2) However, statistical results are valid on their own, without 'hidden' propensities. Popper should not be satisfied with turning propensities into metaphysical ideas. A metaphysical idea does not solve, contra Simkin (1993: 73), the 'mystery of relative frequencies, tending to reach stable values': if a situation has a propensity 'to generate, upon repetition, a certain statistical average in respect of the event', the question remains how this 'mystery' occurs. There is some vacillation in Popper's thinking. On the one hand he affirms that a

> …hypothesis concerning the strength of this physical disposition or tendency or propensity may be tested by statistical tests, that is to say, by observations of relative frequencies. (*Quantum Theory/Schism* 1982/1992: 286)

He repeats this view in various expositions of the propensity interpretation. The propensity interpretation.

> … allows the *single system* to have objective tendencies or propensities of varying strengths or weights or intensities, and shows that, if these are interpreted as tendencies to realize

themselves, they lead to corresponding statistical frequencies in a virtual (or real) aggregate or collective. (*Quantum Theory/Schism* 1982/1992: 130)

On the other hand, the question is how the weights or intensities are to be measured *independently*, i.e. without reference to statistical frequencies?

In *The Self and Its Brain* (1977: 23–4, 30) he points out that sufficient knowledge of the possibility or potentiality hidden in underlying structures would allow predictability of new properties, in principle.

As Popper's enthusiastic response to the results of Eddington's expedition showed, he regarded predictability as a form of testability. What independent evidence exists for propensities? Propensities belong to the 'hidden reality', which Popper increasingly emphasized in his work. Following the quasi-structural view of reality, the question seems to boil down to how structures can be tested. And in this respect Popper's structural interpretation of the laws of nature may provide answers. Popper suggested at one point that propensities may be linked by laws. (*Open Universe* 1982/1988: 95, 105) If propensities can be spelt out in terms of structural properties, then it may indeed be possible to test assertions about propensities, at least in some cases. (Cf. Settle 1974: 740–2)

One such case refers to radioactive decay. E. Rutherford and F. Soddy proposed, in 1902, a 'transmutation theory'.[13] It explained that radioactivity was an atomic phenomenon, which was accompanied by chemical changes. By emitting charged particles—beta particles (electrons) or alpha particles (helium nuclei)—one chemical element was transformed into another. These changes therefore had their seat within the atom and not on the molecular level. The change in the chemical elements was caused by the emission of these subatomic particles. This theory could also explain the observable series of changes in which chemical elements cascaded from longer-lived into shorter-lived elements. But what in turn caused the emission of the particles itself? The radioactive elements, the theory continued, 'must be undergoing *spontaneous* transformation'. (Rutherford/Soddy 1902: 493; italics added) The transformation theory did not stipulate a particular cause inside the atom, which would be responsible for the emission of, say, an alpha particle at a particular time and thus bring about a chemical change. It gives a causal explanation at one level but fails to specify a causal mechanism at a deeper level. In terms of the classical notion of determinism, the transformation theory did not give rise to the precise prediction of emission events. By contrast the theory proposed a formula, which stated the probability value of an alpha- or beta particle emission from a certain element after a certain amount of time. This statistical formula has become known as the *decay law*. These processes can now be explained at a deeper level, involving the weak nuclear force.

In the present context it is not possible to pursue the issue of the testability of propensities and dispositions any further. But given that the propensity interpretation has now become a standard interpretation of probability, the testability of 'tendencies' is a pertinent question.

---

[13] Both scientists received the Nobel Prize for chemistry: Rutherford in 1908 and Soddy in 1921.

Propensities are also a manifestation of indeterminism since they express a disposition, a tendency or the potential to generate frequencies. As such they are irreversible. Once a tendency has materialized, like the decay of an atom, its realization cannot be reversed. The transition from the possible to the actual represents change. 'The idea of propensity distinguishes the unrealized possibilities of the future from their realizations present and past'. (*Quantum Theory/Schism* 1982/1992: 203, 185) Irreversible propensities *also* determine the direction of time. (See *Self/Brain* 1977: 26 [fn4]) Thus, it is indeed the case that the (relational) theory of time was central to Popper's thought.

When Kuhn's influential book on *The Structure of Scientific Revolutions* (1962) appeared, Popper faced a new challenge in his campaign against subjectivism. It was a challenge to his leading idea of testability. Kuhn derived it from his study of the history of science. The dispute with Kuhn prompted him to turn his attention to evolutionary biology. It came in addition to his battle against subjectivism in quantum mechanics, which dated back to Heisenberg's famous 'uncertainty principle' (1927). Furthermore, he was drawn into what he regarded as a fruitless discussion with members of the Frankfurt School of Sociology.

Chapter 6
Popper's Life in England (1960–1970)

The universe now appears to be not a collection of things, but an interacting set of events or processes…. (K. Popper/J. Eccles, *The Self and its Brain* 1977: 7)

In the decade from 1960 to his retirement from the *LSE* (1969) Popper remained as productive as ever, both elaborating on his earlier work and exploring new areas. He was by now an established figure whose work attracted controversy and assent in equal measure. And he was in demand. Feigl, for instance, arranged for him to visit the Minnesota Center from 1 September–31 December 1962 [294.06]. The formal invitation was issued on October 17, 1961. He was offered $7,500 for the four-month period and was expected to 'remain in residence at Minneapolis', barring one or two 'brief leaves of absence'. Popper had been keen to accept the invitation but when he saw the conditions of the official appointment he declined. He could not accept the limitations on his freedom, and he suspected that he would earn less than other visitors before him (10/03/62). Berkeley paid him $9,450 for a three-month period (05/02–15/05/62) (letter 19/03/62). Feigl moved heaven and earth to make him change his mind. The travel conditions could be relaxed, and he was applying to the National Science Foundation to increase his salary to $9,500. Even on $7,500 he would be the highest earner ($1,875 per month). Feyerabend had earned $7,000 for five months ($1,400 per month) and Grünbaum $5,252 for four months ($1,303 per month) (15/03/62). When Feigl's application to the National Science Foundation was approved, Popper was offered $9,500 ($2,374 per month). It was a generous offer but he had to pay taxes, travel expenses, superannuation contributions and of course living expenses. Popper finally accepted and accommodation was to be arranged. Hennie spelt out their priorities (09/06/62): (1) A quiet house in a quiet district. If it only had one bedroom, it should have twin beds, not a double bed. (2) A comfortable living room and a large table. (3) An efficient central heating. It had to be warm throughout the house, including the bedroom. 'The ideal would be a small house of, say, three rooms, in a quiet district, southern exposure, surrounded by a small garden (to keep out noise from neighbours and noise and smell (car fumes) from the road)'. She added that she did not like cooking and that an easy-to-clean kitchen would serve

© Springer Nature Switzerland AG 2022
F. Weinert, *Karl Popper*, Springer Biographies,
https://doi.org/10.1007/978-3-031-15424-9_6

her purpose [294.06; underlined in original]. Similar accommodation requirements were made for their visit to Denver in 1966.

Meanwhile, honours continued to roll in: On March 2, 1965, Popper received a knighthood from the Queen of England. The honour came as a surprise to him. His acceptance may also be surprising to us in view of his republican and social democratic credentials. In the same year he became an Honorary Member of the Royal Society of New Zealand, of which he had been an active member from 1937–45. A year later, 1966, he was elected as Foreign Honorary Member of the American Academy of Arts and Sciences. But health problems continued to plague him.

6.1 Popper at the LSE

In 1960 Popper underwent an operation for the detachment of the retina on both eyes at a hospital in Vienna. There were fears that he would become blind. (*Open Universe* 1982/1988: Foreword; *Autobiography* 1974: 108) Rumours even circulated that Popper had died (as von Hayek informed him on 27/02/60 [305.15]). But none of these frailties slowed down the pace of his work. His fame meant that he and Hennie travelled the world. For instance on May 11, 1963, he described his travel plans to Schilpp [334.02]: they would leave Indiana University on May 14, travelling to Australia, via Honolulu, Tokio and Hong Kong. After their two-month stay in Australia (from May 30 to the middle of August) they would go on to New Zealand, Australia, Saigon, Bangkok, India, Teheran and Egypt. They expected to be back in the UK around the end of August. He paid several visits to the USA, including Brandeis, Denver, Harvard, Princeton and the University of Michigan. He went to Denver at the instigation of Yourgrau, a former assistant of Schrödinger's in Berlin (1930–1932) and recipient of the Einstein medal (1970). Interestingly, Denver wanted him to lecture on topics in the philosophy of the social sciences, a request, which Popper willingly accepted. In preparation for his visit, Denver wished to produce a brochure with photographs, and print 10,000–20,000 copies. Popper was appalled: He wrote to Yourgrau (16/10/65) that 'the idea of a brochure (about me) fills me with horror'. As he did not want to be a killjoy ('Spaßverderber'), he gave in but urged Yourgrau to print as few copies as possible [364.11; my translation]. Before they arrived the Poppers holidayed for a week in Bermuda (22–29/03/66). They spent the Spring Term in Denver (30/03–10/06/66) and Karl received a remuneration of $10,000. He also attended conferences in Geneva (1962), Amsterdam (1967), Salzburg and Vienna (1968). (See Brudny 2002: Chap. 12) The papers he delivered were published in *Conjectures and Refutations* and other volumes. (Popper dedicated his 1963 volume of essays *Conjectures and Refutations* to von Hayek.)

In 1961 Popper was invited to deliver the Herbert Spencer Lecture: 'Evolution and the Tree of Knowledge'. (*Objective Knowledge* 1972: Chap. 7) A few years later, in the second Arthur Holly Compton Memorial Lecture (Washington University 1965, reprinted in *Objective Knowledge* 1972: Chap. 6) he elaborated his ideas. In this famous lecture, entitled 'Of Clouds and Clocks', he proposed what he regarded as an improvement of Darwin's evolutionary theory. It became known as the *Spearhead* model of evolution. It was meant to solve the problem of 'orthogenetic trends'. (That is, Popper tried to tackle the question of the evolution of complicated organs, like the eye.) (*Objective Knowledge* 1972: Chaps. 6, 7) These reflections on evolutionary theory also helped him to formulate his response to Kuhn's views on the history of science.

Further forays into evolutionary biology would follow.[1] His Romanes lecture 'On the Problem of Body and Mind' at the University of Oxford (31/10/1972) was considered a success. In 1977 he delivered a lecture on 'Natural Selection and the Emergence of Mind' at Darwin College in Cambridge. On August 8, 1985, the Royal Society invited Popper to deliver the first Medawar Memorial Lecture, to be given in June 1986. The Lecture was established in honour of Sir Peter Medawar, at the request of an anonymous donor. The Lecturer received a 'voluntary payment' of £500. Popper accepted (29/09/85) but the invitation came at a very difficult time in his personal life. His wife was seriously ill and he had to take her to Wiesbaden, Germany, for treatment. He also decided to change his topic from 'Ideologies as a threat to science' to 'A new interpretation of Darwinism'. In this lecture, as we shall see, he challenged core assumptions of molecular biology. Although the Society asked Popper repeatedly to provide a typescript of the lecture, with the intention of publishing it, Popper failed to do so [561.01]. It was finally published in 2014.

From his home in Penn Popper travelled into Central London to fulfil his duties as professor of philosophy. He looked after doctoral students, such as A. I. Sabra, W. W. Bartley III, and I. Jarvie. Abdelhamid Sabra, originally from Egypt, became professor of the history of science. He taught at the University of Alexandria (1955–62), at the Warburg Institute (1962–72), and at Harvard University from 1972 until he retired in 1996. William Bartley was professor of philosophy at California State University and had begun to write his biography of Popper ('Rehearsing a Revolution') when he died suddenly at the age of 55 (05/02/90; Watkins published an obituary in *The Telegraph*). Jarvie is still the editor of the influential journal *Philosophy of the Social Sciences*. He described his work as Popper's PhD student 'as a wonderful experience'. (Personal email communication 20/02/2019) Popper also came to the *LSE* for the Tuesday seminars, which he chaired with an iron fist. This weekly seminar became famous for its intimidating atmosphere. (See Agassi 2008: 69; Brudny 2002: 173) He was very demanding and, as Watkins recalled, would make devastating interventions. It was a unique achievement when a speaker was not even interrupted once [360.02]. Yourgrau was such an exception: when he gave a talk at the Seminar in 1959 on the nature of physical constants, he thanked Popper afterwards (08/03/59) for his 'positive attitude' [568.06; my translation]. It is worth mentioning that Popper continued to lead this seminar after his retirement until he fell out with Lakatos. But the department continued to call it the Popper Seminar.

---

[1] Although Peter Medawar was an enthusiastic supporter of Popper's hypothetico-deductive method, he criticized him for his 'crude conception of neo-Darwinism' (letters of 29th March 1965 and 16th August 1966, [325.26]). To Monod Popper admitted that his 'excursions into biology' were 'amateurish' (01/09/73) [329.33].

There had been friction between the two men before. In a series of letters, in 1962, both Karl and Hennie complained that Lakatos's manuscripts were typed before Popper's. Karl insisted that his work should be given priority. Hennie penned a long letter from Berkeley (09/05/62) accusing Lakatos of interfering with Karl's work and 'misusing his kindness'. If Hennie had to do the typing, which was the job of the *LSE* secretary, she would get distracted from her and Karl's focus on the *Postscript*. Lakatos defended himself (03/06/62) and stressed how important Karl's friendship was to him [318.05]. A year later this dispute was forgotten. Lakatos started preparations for a Colloquium, to be held at Bedford College in July 1965, and congratulated Popper on his knighthood (12/01/65) [318.07]. But in 1972 another disagreement erupted. In his contribution to the Proceedings of the Colloquium Lakatos quoted—without Popper's permission, as he repeated to Magee [322.16]—from the galley proofs of *Postscript*. (Lakatos/Musgrave 1970: 183) Three times Popper asked Lakatos to return his copy of the galley proofs but Lakatos claimed them legally as his property. In one of his last letters to Lakatos (24/11/72) Popper confessed that he did not 'enjoy disagreements' [318.10]. Watkins tried to tempt Popper back 'to your old Seminar' for the meetings of October 16 or 23, 1973, as he knew that Lakatos would be away in America. Popper answered that he could not attend: 'If Lakatos is there I do not wish to come, because he has been very unpleasant. If he is away, I cannot come, because I do not wish to avoid him' (letters dated 03/10/73; 07/10/73 [567.06]). Although the Department invited Popper to resume his role (as seminar chair) after Lakatos's death, he declined.

Lakatos provoked Popper's displeasure when he criticized him for conflating 'naïve and sophisticated falsificationism'. Popper's reaction to Lakatos's attack was reminiscent of his response to similar criticisms. Lakatos—first Popper's student, then his colleague at the *LSE* from 1960—had 'misunderstood my theory'. (*Autobiography* 1974: 999) He regarded Lakatos's contribution to the Schilpp volume as 'disastrously bad and confused' (28/10/70). His disappointment with his former colleague, whom he had admired, was palpable. Apart from feeling misunderstood, 'I was also surprised by the aggressive and even arrogant tone of certain parts of his lively paper (…)'. ('Replies' 1974: 1000) He was so shocked, as he confided in his former research assistant Alan Musgrave (23/04/71), that he never wanted to see Lakatos again [330.15]. Up to this point the two men had been on good terms. In letters, which Lakatos wrote to Popper in 1959, he praised him as 'a great teacher'. He called himself a 'Popperian' and gave lectures on 'Popperian philosophy' [318.06]. He heralded the forthcoming publications of *Logic* and *Postscript* as 'memorable events in the history of philosophy' [548.06]. Popper told Watkins (18/02/61) that he had a lot of sympathy for Lakatos, a 'wonderful tutor', even though 'he had constantly let me down' [359.34]. In 1964–5 Popper recommended Lakatos for a Readership at the *LSE* as well as a job in Oxford [318.06, 318.07; 270.35]. In 1970 he proposed him for the editorship of the *British Journal for the Philosophy of Science* [330.15]. He even wrote to Ian Jarvie (29/01/75) that he would have formulated his response to Lakatos differently, had he known that 'Lakatos would not survive' him [313.06]. Popper informed David Miller, on February 3, 1974, of Lakatos's unexpected death the previous day. His letter only expressed regret without a tinge of bitterness [328.02]. When asked by Watkins, he agreed to act as a signatory for an Appeal Fund in memory of Imre Lakatos (02/04/74) [360.01]. Maybe this wound would have healed over time, as it did with Bartley.

After von Hayek's move to Chicago, in 1950, Popper complained to von Hayek that he 'had no friends at the *LSE*' and had become 'highly dissatisfied' with his teaching'. (Watkins 1997b: 667–8) However, his feelings changed over time. Ten years later (31/12/60), he told von Hayek that he 'continued to be pleased with my staff here, and my students, although, of course, really good students are always rare' [305.15]. In the same letter he informed his correspondent that his former student Agassi had moved to Hong Kong. He did not stay long and later taught at the University of Illinois, Boston University, York University (Canada) and the University of Tel Aviv. Popper's relationship with Agassi evolved in ways similar to that with Bartley. Their correspondence from the 1950s was friendly and collegiate. But already in 1954 (16/12) Agassi apologized to Popper for being 'too cheeky'. When Agassi was being considered for the position of Head of Department in Hong Kong, he wrote to Popper that the current Head was opposed to his appointment because he was seen as 'argumentative', 'controversial' and 'irreligious' (04/07/61) [266.25]. Yet Popper supplied him and his wife Judith with references. But in 1964 a dispute erupted between them about the mutual acceptance of criticism. Popper complained to Agassi (12/05/64) that it was hard to criticize him, while he, Popper, accepted his criticism willingly. Referring to a particular paper, Popper compared Agassi's statements to 'certain pseudo-arguments of Hegel'. In his reply (15/05/64) Agassi voiced his shock at this comparison and spoke of the 'end of intellectual relations' between them: 'this is a farewell letter' [266.26]. Of course it was not. Agassi continued to send material to Popper and asked Hennie to 'straighten things out' between them. This affair dragged on for several years and eventually suggestions emerged to 'let bygones be bygones'. Popper summed up the situation in a letter (09/05/70) to the science editor of the University of Chicago Press:

> Agassi was my student and we worked closely together for years; but as it sometimes happens, it became important for him to free himself of my (apparently) somewhat oppressive influence [267.01].

Agassi wished to remain friends with Popper on a personal, not on a philosophical level. From Agassi's perspective, Popper fell out with him, because 'he would not accept criticism from any former student'. (Agassi 2008: Chap. I, Prelude) Popper did not see it that way. He thought highly of Agassi, as he told Feigl (23/03/60): 'I know few people to whose opinions I should listen more carefully, especially where it clashes with my own opinions, as happens quite often in his case' [294.06] Eventually things calmed down between them, that is until 1973 when Agassi sent Popper (20/08/73) a copy of his devastating review of *Objective Knowledge* and asked Popper for feedback. It was published as a critical study but it is more of a rambling than a critical review. (Agassi 1974) It is impossible to do justice to these forty pages in a few sentences. I will quote a few passages, which upset Popper. Agassi professes to be a Popperian but he starts with the thesis that 'Popper has regrettably let pass the chance to erect his definitive philosophy'. He proceeds by claiming that

Popper's book projects an author's image that 'is in surprising agreement with the image one gets from philistine gossip'. Agassi hastens to add that this public image is unfair. Still, he finds that Popper's book lacks a clearly stated problem, that his theory of 'explanation is false' (p. 177); that 'his current work, the present volume in particular, (is) quite outdated'; that the 'present volume reverts to old positivism', and that 'his distinction' (between World 1, 2, 3) 'is trite' (p. 190). Popper was naturally scandalized by this review and for many years did not want to receive any more communications from Agassi. Agassi, however, continued to write to him but his advances fell on deaf ears. It was only when Agassi sent him a general apology (28/07/92) as a 'peace offering' that Popper finally relented:

> Dear Joske, Thank you for your Fax dated "1992.07.28". I accept it gladly, as "a peace offering", as you put it (01.08.92) [526.08].

In his last decade at the *LSE* tensions also arose between Popper and Watkins. Watkins, however, remained supportive of Popper's ideas, especially his *Spearhead Model* of evolutionary theory. Not everybody felt that way. He attracted hostility and admiration but not in equal measure. Both David Miller and Ian Jarvie told me that they got on well with Popper. He was kind, generous, had a good sense of humour but was very demanding. This verdict is reflected in his correspondence with his loyal friends. Helmut Schmidt experienced him as a 'modest man in a modest house' (Schmidt 1987). The journalist Leslie White, who interviewed him at the end of his life, also portrayed Popper as 'a modest man by nature and principle' (White 1992).

Popper liked to give his assistants very precise instructions about how to handle their jobs. David Miller described to me his work as an assistant (personal communication 09/04/2019):

> A typical working day began at 10.00, when I called him at home, and he would mention some books and papers that he would like to read. I then spent some time running these to earth in the LSE Library, in other London libraries, and beyond, and posting them to him. During term-time, he usually returned the books by hand when he came into the School. He liked it that I set up a simple system that kept track of why a book or a paper (on occasion not obtained until several weeks after the search began) was required. (…) Popper's hand-written MSs were typed at home by Lady Popper, and then sent to me. I would have them retyped (with several carbon copies) by his secretary in the School, and I would then enter comments and criticisms (which he nearly always took seriously). This cycle could be repeated several times before a typescript was sent to an editor.

In addition, Miller, like the other assistants, was responsible for forwarding mail to Popper, when he was abroad. But their correspondence also bristled with philosophical discussions about probabilistic support, verisimilitude and the impossibility of inductive logic. Popper would send Miller manuscripts and proofs for corrections. He praised him as an excellent research assistant and was instrumental in Miller's appointment at Warwick [327.14–15]. The two men got on so well that in 1975 Popper donated wardrobes to Miller's new household and ten years later he offered him money for a deposit on a new house [328.02; 328.06]. In 1981, Miller started preparing his edition of *A Pocket Popper*. It was published in 1983 and sold about 1,500 copies in the following decade [552.04].

Of immediate concern in the early 1960s were attacks on his methodology and his views on the social sciences. These disagreements targeted both his 'falsificationism' and his objectivism. Since his association with the *Vienna Circle* a 'legend' had spread that Popper was a positivist. When he was struggling to find a publisher for his first book, he wrote to J. Kraft (25/05/33 and an undated letter probably from 1932) that 'he did not feel like being labelled as a positivist'. Even Carnap seemed to make a 'staunch positivist' out of him [Hansen 3.2; my translation]. Reichenbach opened his review of Popper's *Logik* (1935) with the observation that Popper 'unfolds his thoughts in close analogy to mathematical–physical questions and by employing terminology, which has been developed by logical positivism' [Hansen 3.2; my translation]. Now, his apparent positivism became the battle cry of prominent members of the Frankfurt School of Sociology. On both fronts Popper was called upon to defend his objectivist stance. It was a fight against the encroachment of subjectivism—or so he saw it— which took him back to physics and his opposition to holism. It also led to the reaffirmation of the open society.

It is revealing that in his *Autobiography* (1974) Popper omits to mention two of his influential critics whom he regarded as post-rationalists and post-critical thinkers.[2] One was Michael Polanyi, a British-Hungarian polymath, with whom Popper had corresponded since 1932 [339.01]. At Polanyi's invitation Popper went to Manchester, in June 1946, to give a talk on the topic of the open society. Polanyi expected to get involved in a controversy with Popper (04/05/65) [339.01]. To philosophers Polanyi is especially known for his book *Personal Knowledge* (1958). The other critic was Thomas S. Kuhn, whose famous book *The Structure of Scientific Revolutions* enjoyed several editions. Kuhn's work came to dominate the philosophy of science for decades. His notion of 'paradigms' took hold in the social sciences and infiltrated political and social discourse. Popper did respond to Kuhn on numerous other occasions. It can be said, I think without exaggeration, that Kuhn's ideas overshadowed Popper's work. So did Lakatos's in the 1970s. In contrast to Polanyi and Kuhn, Lakatos is mentioned in the *Autobiography*.

His debates with the post-rationalists and the left-wing sociologists of the Frankfurt School took place within a particular context. Some of his critics were reacting to Popper's views on falsifiability in science, whilst others objected to his insistence on piecemeal social engineering. His response to these criticisms led to a defence and modification of his earlier views. Let's first consider the challenge from the Frankfurt School and then the more influential views of the post-rationalists.

---

[2] Recall that Popper's fight for objectivity was not a reaction to Polanyi's or Kuhn's attacks. It started much earlier, namely with his interest in quantum mechanics, especially Heisenberg's uncertainty relations, and in indeterminism, which led to his propensity interpretation of probability. (See *Logic* 1959/1980: §8; Postscript Vols. I, III; 'Replies' 1974: 1067)

6.2 Popper's Debate with the Frankfurt School of Sociology

In 1960 Popper was invited to give the opening address at a congress of German sociologists in Tübingen. The meeting took place in October 1961 under the title 'The Logic of the Social Sciences'. Theodor Adorno was to reply to Popper's theses from the perspective of the Frankfurt School. By his own admission Popper had not been fully aware 'of the influence of the Frankfurt School' when the invitation arrived. (Letter to the Editor, *Times Literary Supplement* 26/03/1970) Prior to the meeting a friendly and respectful correspondence took place between Adorno and Popper. They exchanged notes and papers, assuring each other of the many agreements between them. They even concurred that the main task of sociology was criticism of society (This correspondence is published in Braunstein 2019). Yet, during the meeting, Popper felt that speakers were talking at cross purposes, an impression which was confirmed when the talks were published under the title *Der Positivismusstreit in der deutschen Soziologie* (1969).[3] The title of the book *The Positivist Dispute* was misleading, since it cast Popper as a positivist. This myth, which he tried to debunk on numerous occasions,[4] arose because of the publishing history of his first book *Logik*. As mentioned above, the book appeared in a book series that was edited by members of the *Vienna Circle*. Popper also reacted very sharply against his opponents' abstract terminology. Their 'verbiage', which Popper had already condemned in Hegel, seemed to hide the shallowness of their ideas. They were influenced by the Enlightenment, as well as Hegel, Marx and Weber. But on another level, his disagreement with the Frankfurt School centred on the contrast of reform *versus* revolution, piecemeal engineering *versus* radical change.[5] Apart from Theodor Adorno, the most influential members of the School were Max Horkheimer, Herbert Marcuse and the younger Jürgen Habermas.

---

[3] English translation: *The Positivist Dispute in German Sociology* (1976). The French title of the book is less controversial: *De Vienne à Frankfort: La querelle allemande des sciences sociales* (1979).

[4] *Autobiography* (1974): §7; 'Replies' 1974: 963–74; *Auf der Suche* (1987): I.6; *Myth* 1994: Chap. 3; *Alles Leben* 1996: 162, 248; *All Life* 1999: 85.

[5] Popper's insistence on piecemeal reform prompted the British historian Carr (1964: 155) to complain: 'The status of reason in Professor Popper's scheme of things is, in fact, rather like that of a British civil servant, qualified to administer the policies of the government in power and even to suggest practical improvements to make them work better, but not to question their fundamental presuppositions or ultimate purposes.' By contrast Schödinger was very impressed with piecemeal engineering and told Popper (04/08/49) that it also occurred in nature [347.18].

Max Horkheimer (left) greets Theodor Adorno (right). Jürgen Habermas brushes back his hair, on the right in the background. *Source* Wikimedia Commons

They developed what is known as *Critical Theory*. In order to assess the problem situation that Popper was facing, it is only fair to briefly characterize the main tenets of Popper's opponents—first the proponents of *Critical Theory* and then the post-rationalists. *Critical Theory* defended the following ideas:

- Social Science cannot treat society as an undetached object of research. The social scientist is inevitably engaged with the object of research—society—and must use her/his findings to criticize it. This does not mean that the methods of research are biased. But the research results should be used to reveal the actual state of society.

- Scientific rationality should not be the paradigm of rationality in society, since it is only a means-end rationality; i.e. it does not allow the social scientist to question the state of affairs in society. Weber identified this kind of rationality as purposive-rational action.

- An empirical approach to the social sciences only occupies itself with instrumental rationality. Instead of criticizing society it only states facts. But the social sciences should preserve a fuller sense of rationality, as laid down in the Enlightenment. It is a type of rationality that not only provides means to external ends, but also ways of judging the rationality of ends. The Enlightenment philosophers did lambast the very foundation of eighteenth-century society: the entanglement of Church and State, absolute monarchy, censorship of the press, the blind obedience to authority and tradition.

- Social science should not just provide a description of society as it is. It should also propose a blueprint of what it *could* and *should* be; i.e. the distinction between facts and norms should be questioned.

In the course of its development *Critical Theory* combined a criticism of scientific rationality, which it dismissed as instrumental, with a reflection on the methodology of the social sciences. A book entitled *Dialektik der Aufklärung* (written in 1944 but first published in 1947, translated into English as *Dialectic of Enlightenment*) became the 'manifesto' of *Critical Theory*. Its authors were Max Horkheimer and

Theodor Adorno. It is a difficult book to read. It is easier to follow Horkheimer's monograph *The Eclipse of Reason* (1947), which provides a good introduction to *Critical Theory*. In the Preface Horkheimer (1974: vi) writes:

> As understood and practiced in our civilization, progressive rationalization tends to obliterate that very substance of reason in the name of which this progress is espoused.

The basic thesis of *Critical Theory* is that the concept of rationality underlying modern industrial societies has destroyed the very aim—the emancipation of humankind—from which it originated. Or the primary aim of scientific rationality, as developed by the Enlightenment philosophers, has backfired: objective reason has been reduced to instrumental reason. The Enlightenment philosophers had aimed at a global reform of society. Their argument was: if nature can be understood rationally, as the Scientific Revolution had proved, then society should be understood rationally, too. This approach resulted in the belief that society was subject to 'societal laws', which the social sciences should endeavour to discover. Horkheimer identified the 'present' crisis of reason as its formalization. Reason was reduced to a mere instrument in the service of an external end. Objective reason, by contrast, should incorporate a reflection of ends and their desirability. If scientific reason is nothing but a means for an appropriate end, the end itself must be determined by some arbitrary decision. It is set by an external agent: economic or political interests. What this means, according to Horkheimer, is that instrumental reason tends to become independent and suppress its former companion, i.e. the rationality of ends, as he detected it in the European Enlightenment. The members of the Frankfurt School shared the view that advances in technical facilities (as an outcome of scientific research) are not to be identified with social progress. What philosophy (or critical sociology) can do is to function as a corrective by analysing the present state of reason in industrial societies. An excessive use of instrumental reason could drive society to the brink of destruction. This philosophical analysis, if translated into praxis, would mean that we require 'a definition of the objective goals of society including the respect for individual life'. (1974: 175) Horkheimer ends his essay with the words:

> …the denunciation of what is currently called reason is the greatest service reason can render. (1974: 187)

Which implications does this conception have for the social sciences? The proponents of *Critical Theory* drew a distinction between a *critical* and a *conservative* function of the social sciences. Conservative functions are restricted to the use of instrumental reason. Critical functions preserve an interest in the progress of society as a whole. Hence critical sociology retains an engagement with the rationality of ends, to which society should be committed. These ends include interest in (social) justice, emancipation, fairness and the fulfilment of individual potentials. Today Habermas sees the function of critical social research in the study of the paradoxes and pathologies of societal rationalization, i.e. the confusion between increase in complexity and increase in rationality. (Habermas 1987, Volume II: VIII.3)

Critical sociology does not depart from traditional sociology in its application of objective methods. But it does differ from it in its interpretation of *how* the empirical results of the social sciences are to be employed. Critical sociology holds that neutrality towards the implementation of empirical results leads to irrationality in the area of decision-making. The results of social research should not only be used to describe how society *is*, but also how it *could* and *should* be. Sociology should be committed to an ideal—let us call it a fair, free and emancipated society—so that the actual state of society, described by objective empirical methods, can then be compared to that ideal. It is important to realize that according to Horkheimer and Adorno the technological means exist today to 'make the world a better place'. But the ends society pursues are not 'rational' from their point of view. The tragedy of modern society is the clash between the availability of means to achieve the dreams of the Enlightenment, and the actual state of world affairs. And they blamed the state of affairs (today: arms race, climate change, overpopulation, pollution, poverty) on the irrationality of ends, due to the instrumentalization of reason. Popper shared some of these concerns.

The problem, as the reader will surmise and as Horkheimer himself admitted, is how to transform this philosophy, i.e. *Critical Theory*, into concrete social research. (Horkheimer 1970: 54; 1972: 239) The proponents of *Critical Theory* made some attempts in this direction. Their empirical studies focussed on authoritarianism, studies of the effects of mass media, reflection on scientific methodology and the socialization process. (Adorno and collaborators published studies in Prejudice, including a volume on *The Authoritarian Character*, 1950; Habermas criticized instrumental reason and positivism in the social sciences and developed his *Theory of Communicative Action* 1984; 1987).

It was to be expected that such a characterization of a critical function of social research would encounter strong resistance. It centred on the value neutrality in science. This postulate states that science is only concerned with what is, not with what should be. From this point of view, the critical conception of social science violates the postulate of value neutrality because it seeks to recombine the severed links between theory and praxis. *Critical Theory* rejected what it saw as the irrational decisions regarding societal objectives. Proponents of the Frankfurt School derided this attitude as *decisionism*. (Habermas 1976: 140–9) This reproach can be made because the social sciences may have an effect on the social processes they investigate. Sociologists study ethnic minorities, television viewing patterns in young children, delinquency in juveniles or underachievement in education. Their findings could lead to a change in societal practices. Traditional social scientists are aware of this effect.

The contrast between the critical and conservative functions is the contrast between wholesale and piecemeal engineering. (Popper, *Poverty* 1957/[2]1960: Chap. 21) Popper compared 'piecemeal social engineering' to

> …physical engineering in regarding the *ends* as beyond the province of technology. All that technology may say about ends is whether or not they are compatible with each other or realizable. (*Poverty* 1957/[2]1960: 64)

This instrumental rationality stands in stark contrast to what Popper dismissed as utopian engineering. (*Poverty* [2]1957/1960: 67) The worry of *Critical Theory* is whether such a voluntary restriction is sufficient, given the basic premise that society is both the *object* and the *subject* of the social sciences. According to Popper, the

piecemeal social engineer looks upon social institutions from a 'functional or instrumental point of view. He will see them as means to certain ends or as convertible to
the service of certain ends.' (*Poverty* 1957/²1960: 65) In the eyes of the Frankfurt
School, Popper endorsed the conservative function of sociology.

Critical Theory suspects precisely that the means to ends relationship is insufficient: What if the ends themselves are irrational? Should a social scientist suggest
means to an end s/he finds unjust or objectionable? This worry boils down to the old
question of value neutrality, as discussed by Weber. Popper insisted on this value
neutrality: natural and social science have to separate the values *internal* to scientific research—coherence, explanatory power, fruitfulness, precision, search for truth,
simplicity, testability—from values *external* to this research, i.e. the reform of society,
the emancipation of humankind, etc. (Popper 1976: 96–8) But these are precisely
the values which belong to the encompassing concept of rationality, favoured by the
Frankfurt School. In an afterthought to the debate Popper agreed with 'Horkheimer's
vision of a society of free men.' (*Myth* 1994: 80) The problem is that both conservative and critical sociologists may agree on the ideal of an emancipated society but
disagree on how to implement it.

Popper's empirical model	Critical theory model
Piecemeal social engineering	Global reform of society
Separation of the appropriate means from the aims; particular aims are subjected to the trial-and-error method	Global aims/ends should be part of a rational debate
Objectivity of research	Objectivity of research
Value neutrality: exclusion of external values from scientific work	Reform of society as an objective of scientific research

What was wrong, according to Popper, with such a blend of internal and external
values in the social sciences?

The more ambitious programme, associated with *Critical Theory*, runs into the
same notorious problem as Marxism, i.e. the problem of unplanned planning. Popper
dubbed it the *Oedipus* effect. (*Poverty* 1957/²1960: 13) It describes the effect of a
prediction on predicted event. It is one of the central aspects of the relationship
between the social sciences and their object matter, i.e. society, that the interference
with a social process may change its direction, as illustrated in self-fulfilling prophecies (food prices will rise) and self-negating prophecies (stock markets will fall).
(See *Poverty* 1957/²1960: 12–4) Hence, if a global goal were to be pursued—such
as the emancipation of society—unplanned situations would arise. An infinite circle
would ensue because of

(a) the uncertainty of human agency,
(b) the uncertainty of predictions in the social sciences, due to the complicated
      interaction between observer and observed (*Poverty* 1957/²1960: 12–4, 65–70;
      cf. Elster 1989: 168),
(c) the lack of societal laws.

Did Popper's argument throw the critical conception of sociology into an irresolvable dilemma? Adorno and Horkheimer remained committed to the global concept of rationality. Adorno argues that 'critical sociology is necessarily a critique of society'. (Adorno 1976: 114) Horkheimer agrees.

> Critical theory never aims simply at an increase of knowledge as such. Its goal is man's emancipation from slavery.

And he continues:

> The new dialectical philosophy, however, has held on to the realization that the free development of individuals depends on the rational constitution of society. In radically analyzing present social conditions it became a critique of the economy. (Horkheimer 1970: 58; 1972: 246)

The important idea, from Adorno, Horkheimer to Habermas is that

> …(i)t is impossible to understand social components without referring back to the whole of society. To understand individual social processes and how they function, it is necessary to refer back to the whole society in which they are embedded. (Adorno 1976: 105-7; cf. Habermas 1976: 131-2)

If *Critical Theory* envisages an overall direction and a goal for society, Popper's criticism has force. But there is a difference between saying (1) that *Critical Theory* harbours an interest in emancipation and claiming (2) that *Critical Theory* aims at a global reform of society: an emancipation of humanity from unjust social and economic circumstances. The early *Critical Theory* was rooted in Marxist political thought and saw its ultimate aim in the destruction of class society.

But there is a different sense of emancipation, which Popper's argument does not touch in this context. It goes back to the original sense of the Enlightenment as 'illumination': shining a light on a state of affairs and providing information about it. Popper invoked this sense in his correspondence with Isaiah Berlin. In this more sophisticated version, *Critical Theory* accepts that scientific rationality is the life form in a modern society. What it requires is Enlightenment about its consequences. According to Adorno and Horkheimer rationality had turned into instrumental reason to such an extent that it had swallowed its own reflexive capacity. They characterized instrumental reason as a machine churning out technological recommendations to whatever end was provided. Habermas, however, stressed that scientific rationality included a critical potential. His aim was to illuminate the very presuppositions of scientific rationality; and one of them was that scientific rationality is guided by technological interests. Insofar as the social sciences follow the empirical model they too pursue technological interests. And the danger is that practical concerns are simply identified with technological issues. This is how Hans Albert (1976: 163–6), a defender of Popper's stance, characterized Habermas's position.

The (sociological) Enlightenment Habermas envisages is the realization that instrumental rationality will engender a technological civilization. In other words, it is the awareness that an increase of material knowledge, technological spin-offs and dominion over nature restrict the scope of rationality. This, according to Habermas, is Enlightenment about the sort of scientific rationality that has developed. Habermas's concern is to lay the foundations of a critical theory of society. It is grounded in a comprehensive theory of rationality, which preserves the emancipatory aspirations of the Enlightenment. In this project instrumental reason is only one aspect of rationality. Rationality in the global sense is now understood as communicative rationality. Communicative rationality aims at seeking consensus through the use of arguments and the critical assessment of validity claims. (Habermas 1981/⁴1987: 28, 513–8; 1984: Chap. IV.2) This position is close to Popper's emphasis on the argumentative function of language.[6]

In his response to the proponents of *Critical Theory* Popper did not reflect on this emancipatory aspect of the Enlightenment, although he shared it. He engaged with the members of the Frankfurt School only on two occasions. At the Tübingen meeting, Popper presented 24 theses, which partly repeated positions he had already developed in his earlier work. His main thesis was that both the natural and the social sciences apply the trial-and-error method. He insisted on the value neutrality of research, although he admitted that this is a value in itself. He accepted that external values exist but held that they have to be kept apart from the search for truth and the respect of internal values. He also re-emphasized the autonomy of sociology and his opposition to psychologism. With particular reference to the social sciences he regarded their main task as the description of the social world with the help of explanatory theories. (Popper 1976: 101; *Auf der Suche* ²1987: 203; *In Search* 1992: 180) It is interesting that Popper appealed to the notion of *situational logic*, which he would again employ in his debate with Kuhn. The method of situational logic is not based on the psychological exercise of empathetic understanding. Rather it describes objective understanding: to explain human behaviour in the context of the situation in which it arises. The question is whether the behaviour adopted by social agents is appropriate for the particular situation in which it is displayed. Once again Popper made use of Weber's ideal–typical models, since he held that the reconstruction of human behaviour within a particular situation is subject to empirical tests and criticism. The tests concern the adequacy of the reconstruction, the model, in the face of the real behaviour of the social agent. It may lead to improvements in the objective understanding, which the social scientist gained of the situation under study.

Popper did not really address the issues raised by Adorno and Horkheimer, that is that instrumental rationality may pose problems. When an edited version of the debates was published first in German (1969), then in English (1976), Popper vented his grievances. He rejected the label of neo-positivism with which Adorno and Habermas had tainted him. He complained about the 'big words' used by the members of the Frankfurt School. (*Auf der Suche* ²1987/*In Search* 1992: I. 6) He bemoaned that his contribution was drowned in words. He had avoided discussions with Habermas or Adorno because he objected to the 'cruel game to express what is simple in complicated terms and what is trivial in difficult terms'. (*Auf der Suche*

---

[6] It is certainly true that Popper had a much better understanding of the natural science than Habermas. But both maintain an interest in Enlightenment, the role of rational debate in the public sphere and a dislike of instrumental reason, as Shearmur (1996: 164–68) and Hacohen (2000: 4–5) observed.

[2]1987: 112; *In Search* 1992: 94) Popper was tired of being called 'a positivist'. He claimed that he never took the Frankfurt School seriously.

Although Popper complained about Adorno's verbosity, Horkheimer's vacuity and Habermas's incomprehensibility his real target, it seems to me, was the global reform of society, as proclaimed by (at least) Adorno and Horkheimer. (*Myth* 1994: 78–81) In the social field his concern was with practical issues. His *social theory*—a gradual, piecemeal reform of society and its institutions—contrasted with his '*theory of method*, which happens to be a theory of scientific and intellectual revolution.' (*Myth* 1994: 68; italics in original)

In his later years Popper increasingly talked about scientific revolutions. By and large such talk had been absent from his earlier work.[7] It was no doubt a reaction to the influence of the post-rationalists, in particular the work of Thomas S. Kuhn. But, unlike the latter, he treated scientific revolutions as rational affairs. In this decade, the question of the growth of scientific knowledge and its rational character came into sharp focus. And Popper began to discuss progress in science from the point of view of natural selection more systematically. (*Myth* 1994: Chap. 1, §2) Applying Popper's situational logic to the evolution of his own ideas reveals the context, in which this emphasis on the growth of scientific knowledge occurred. The context was the appearance of what Popper perceived as the rise of post-rationalism.

6.3 Popper and the Post-Rationalists

It will suffice to concentrate on his earlier opponents, Michael Polanyi and Thomas S. Kuhn because Popper responded to their criticism and modified or extended his views. He was disappointed in his later critics, I. Lakatos and P. Feyerabend. He wrote to von Hayek in December 1969 that Feyerabend had 'lost his head completely and has become a leftist Scharfmacher' [agitator] [305.15]. But this statement does not sum up the long-term relationship he had with Feyerabend. Their correspondence ranged from philosophical questions of quantum mechanics, Feyerabend's job searches, his marriage plans and health problems to his singing lessons. In 1954 Popper offered Feyerabend a Research Assistantship at the *LSE*, which he accepted. He also wanted him to translate *Logik der Forschung* into English [294.16]. This Feyerabend did not do but he prepared the first German translation of *The Open Society*, which appeared in 1957 (Volume I) and 1958 (Volume II), respectively. Unlike some other German philosophers, Feyerabend was 'non-verbose and non-pompous' (as Popper wrote to Feigl on June 5, 61 [294.06]).

---

[7] However, already in the early 1930s, he had spoken of crisis as a normal situation in science (see Chap. 6.4.1, p. 136). In *Open Society* II (1945/[5]1966: Chap. 11, p. 12) he repeated that science does not grow by a 'gradual encyclopaedic accumulation of essential information, as Aristotle thought, but by a much more revolutionary method; it progresses by bold ideas, by the advancement of new and very strange theories (such as the theory that the earth is not flat or that 'metrical space' is not flat), and by the overthrow of the old ones'.

6.3.1 Michael Polanyi

Michael was the younger brother of Karl Polanyi, the Austro-Hungarian economic historian, with whom Popper had discussions in Vienna in the early 1920s. Karl, whom Popper mentions in his *Autobiography*, is the author of the well-known book *The Great Transformation* (1944). Michael became professor of physical chemistry, first at the *Wilhem Kaiser Institut* in Berlin, then at the University of Manchester (1933), after his and Karl's emigration to England. Even though Kuhn's work became much more influential than Michael Polanyi's, the latter articulated some of the themes, which Kuhn made popular. Polanyi reacted against the idea of 'scientific detachment': objective criteria of verifiability or falsifiability. He referred to skills and implicit rules, which scientists require to do their work. He and Kuhn called these skills 'tacit knowledge'. In view of such skills—personal knowledge and judgement—Polanyi rejected complete objectivity as an illusion. True discovery is not the result of a strictly logical procedure. It requires judgement. Judgement is of course not purely subjective, for it is guided by skill and experience. Polanyi's book has the subtitle: *Towards a Post-Critical Philosophy*. It makes no reference to Kuhn but it is critical of 'falsifiability'. Polanyi considers an area of research, in which Popper's methodology seems out of place. It is the classification of crystals. The example is similar to the above-mentioned research on the shape of pebbles (see Chap. 3.5, p. 41), because both areas seek to order the phenomena.

> Here stands revealed a system of knowledge of immense value for the understanding of experience, to which the conception of falsifiability seems altogether inapplicable. Facts, which are not described by the theory create no difficulty to the theory, for it regards them as irrelevant to itself. Such a theory functions as a comprehensive idiom which consolidates that experience to which it is apposite and leaves unheeded whatever is not comprehended by it. (Polanyi 1958: 47)

Michael Polanyi (1891–1976). *Source* Springer Link

Popper's methodology was of course not concerned with such classifications or attempts to understand why pebbles on the beach have an oval shape. It is part of science but not the 'heroic science' of the greatest scientists on which he concentrated. ('Replies' 1974: 977f)

Polyani's point, however, which Kuhn shared, was that personal knowledge inspired all scientists. Personal knowledge, he claimed, commits the scientist to a vision of reality.

> Of this responsibility we cannot divest ourselves by setting up objective criteria of verifiability – or falsifiability or testability, or what you will. For we live in it in the garment of our skin. (....) Such is the true sense of objectivity in science.... (Polanyi 1958: 64)

Polanyi pleaded for a *personal participation* of the knower in the act of understanding. Unsurprisingly the first part of his book is entitled *The Art of Knowing*, whilst the second part bears the title *The Tacit Component*. This element of appreciation does not render scientific knowledge 'subjective'.

There is one further aspect in which Polanyi anticipated Kuhn. It is the area of scientific controversy. Such controversy arises when two or more theories are in competition, as happens so often in the history of science. If, say, two theories vie for acceptance in the scientific community—as is currently the case with String Theory and its lesser-known rival Loop Quantum Gravity—then logical reasoning is insufficient. The proponents of the rival theories—Kuhnian paradigms—must go beyond arguments and *convert* their opponents by persuasion. The two paradigms (schools) are at odds with each other, which means there is a logical gap between them. As Polanyi puts it:

> They think differently, speak a different language, live in a different world, and at least one of the two schools is excluded for the time being (whether rightly or wrongly) from the community of science. (Polanyi 1958: 151)

With such passages Polanyi emphasized an idea, which came to dominate discussions surrounding Kuhn's work: the notion of *incommensurability*.

> We can see now, also, the great difficulty that may arise in the attempt to persuade others to accept a new idea in science. We have seen that to the extent to which it represents a new way of reasoning, we cannot convince others of it by formal argument, for so long as we argue within their framework, we can never induce them to abandon it. Demonstration must be supplemented, therefore, by forms of persuasion which can induce a conversion. (Polanyi 1958: 151)

Popper's objectivism implied a rejection of the notion of incommensurable theories. He said so explicitly: Theories can always be compared logically, as long as they

> ...have historically grown out of the same problems (say, to explain the movements of the heavenly bodies). (Popper, *Myth* 1994: 54)

Popper hardly paid any attention to the notion of *paradigm*, which became a fashionable term amongst social scientists in the wake of Kuhn's work. He certainly had not used this term in his previous work, although he later seemed to concede that paradigms in science exist. (*Quantum Theory/Schism* 1982/1992: Preface X, p.31) But Kuhn's work really focussed on *scientific revolutions*, an aspect of the history of science to which neither Polanyi nor Popper had given much thought. After the publication of Kuhn's book in 1962 Popper employed the term 'scientific revolution' much more readily than before. Kuhn meant by a scientific revolution a radical change in paradigms, a transformation of worldviews. But Popper, given his belief in the comparability of scientific theories, depicts scientific revolutions as rational affairs:

.... scientific revolutions are rational in the sense that, in principle, it is rationally decidable whether or not a new theory is better than its predecessor. (*Myth* 1994: 12)

On a personal level Popper first met Kuhn at Harvard in 1950 when he gave his William James lectures. Kuhn, who was a student there, attended Popper's seminar. As he wrote to Munz (05/08/70) they remained on friendly terms [330.10]. He met Kuhn again in 1965 at the colloquium at Bedford College and years later at Berkeley where Kuhn by then held a professorship.

6.3.2   Thomas S. Kuhn

Kuhn had already come to the attention of the scientific community through his magisterial work on the *Copernican Revolution* (1957). It is a detailed, technical treatise on the history of astronomy from the Greeks to the seventeenth century. It revealed hardly a hint of what was about to come with the publication of *The Structure of Scientific Revolutions* (first edition 1962). It will be useful to remind the reader of Kuhn's major theses by contrasting the Kuhnian and the Popperian views.

The *first* point of disagreement was over the goal-directedness of scientific growth. Kuhn held that the growth of science is an evolutionary process towards a more refined understanding of nature but not towards a goal, like truth. Kuhn accepted that the list of problems, which science solves, would grow but the process did not involve an approximation to truth. Kuhn's evolutionary view of scientific growth has a Darwinian flavour, Popper's has a Lamarckian flavour.[8] Both Darwin and Kuhn portray the evolution of species and scientific knowledge, respectively, as a unidirectional process; a process towards complexity but not towards truth. Lamarck described evolution as a necessary progression towards an aim, i.e. the emergence of humanity. Popper held that the *aim* of science was truth. The accumulation of knowledge is supposed to take scientists closer to the truth. This aim may forever elude them. Their search for better theories is based on the Darwinian *method* of

---

[8] This characterization seems to contradict Popper's insistence that the method of acquiring knowledge is 'Darwinian rather than Lamarckian.' (*Objective Knowledge* 1974: 149; cf. *Brain/Self* 1977: 459) Popper describes Darwinism as a method based on 'selection' and Lamarckism as 'instruction by repetition'. However, I am referring to the growth of knowledge, not the method of acquiring knowledge. Unlike Lamarck, Darwin did not see evolution as progressive. According to Lamarck, evolution pursues an aim, namely the eventual emergence of human beings as the crown of creation. The ideal aim of science, according to Popper, is truth or more realistically verisimilitude.

trial and error (selection). It is important for Popper to stress that the rational character of science is only guaranteed if its growth is cumulative. (See *C&R*: Chap. 10, §§1, 17) He called this approximation to truth *verisimilitude* (Box 2): approximation occurs through a 'negative feedback, through the discovery of mistakes' (letter to Lorenz 21/01/75, [321.04]; my translation). But he imposed stringent requirements on successive scientific theories to make sure that one theory would always be replaced by a better theory, i.e. one with greater scope, greater explanatory and predictive power. It is worth mentioning in this connection that Popper regarded the growth of Big Science as a threat: it would lead to stagnation. (*Myth* 1994: Chap. 1, §9)

In the Darwinian picture, species fight for survival by adapting to a changing environment. The winners survive and their reward is reproduction. Their 'fitness' is related to their ecological niche, and the process is natural selection. But evolution, in the Darwinian picture (unlike the Lamarckian view), involves no goal. The environment changes in a random fashion and species have to adapt. Apply this Darwinian picture to the growth of scientific ideas, as Kuhn did. Like biological evolution, scientific development is a unidirectional and irreversible process. Theories are related to each other by descent. Later theories are better at solving problems than earlier ones. Each successful new theory abandons old assumptions and establishes new ones. Often it also invents new vocabulary and defines new problems, while it may dismiss old problems as irrelevant. New theories emerge in a different environment to which they are successfully applied. But if the environment changes and science enters a crisis, then a new theory will emerge that has survived as the 'fittest'.

The *second* point of disagreement is on how theories succeed each other. Popper postulated strict requirements discussed below. A theory which fulfils these requirements can be said to enjoy a higher degree of corroboration than its rivals. This leads to a cumulative view of scientific growth. Kuhn rejected this picture. For him scientific theories replace each other as a result of revolutionary changes. The growth of scientific knowledge is discontinuous. The following scenario unfolds:

Scientists have a certain theory at their disposal—Kuhn uses the term 'paradigm'. They work with it; i.e. they refine it, tidy it up and use it to solve certain problems.

But then a particularly disturbing anomaly arises. An anomaly is a persistent non-agreement between experimental results and a theory. (Kuhn 1977: Chap. 8, §4) One example is Galileo's observation of Jupiter moons, which posed a very serious challenge to the geocentric worldview.[9] Why? In Popperian terms, the geocentric theory literally *prohibited* planetary orbits which did not have the earth as their centre. But the Jupiter moons are centred on Jupiter. Another example, close to Popper's heart, is the perihelion advance of Mercury which was known but could only be fully explained by Einstein's General theory of relativity (Fig. 6.4).

---

[9] Further examples of scientific anomalies are (a) the large-scale scattering of electrons for Thomson's plum-pudding model of the atom; (b) the discovery of the eccentric orbits of comets; (c) the age of the earth and the paucity of the fossil records for Darwin's evolutionary theory in *The Origin of Species* (1859).

If an anomaly strikes, the paradigm, according to Kuhn, enters a period of *crisis*. Scientists search for solutions. Eventually a new paradigm gets accepted, which, however, radically changes major assumptions, to the point of introducing a new ontology. Kuhn claims that there is 'no theory-independent way to reconstruct phrases like "really there'''. He considers that there is no match between what a theory claims there is and reality. (Kuhn ^{2}1970: 206) This claim runs counter to Popper's common-sensical view of realism, as the existence of an external world irrespective of human awareness.[10] The new paradigm, according to Kuhn, brings with it revolutionary changes in the whole outlook it offers to the world. The way people see the world is heavily dependent on the paradigm they accept. As paradigms change, so worldviews change. Proponents of different paradigms reside in different worlds. All this shook Popper's deepest intuitions about objectivity. (Later he would place scientific theories in an abstract objective realm, called World 3.) Looking at the history of science, Kuhn became convinced that scientists adopted scientific theories not as intellectual tools to comprehend the world, but as worldviews, as grids, through which they ordered the world. A paradigm becomes a basic mode of thinking, which people cannot easily shake off. It is just the way they perceive the world—through Ptolemaic or Copernican eyes, through Lamarckian or Darwinian lenses—and it requires a revolutionary jolt to alter the way people think about and look at the world. Kuhn's focus was as much on scientists who embrace paradigms as worldviews as on the worldviews themselves. Scientific knowledge becomes a property of a group of scientists. By contrast, Popper opposed the thesis that scientists must believe in their theories. (*Auf der Suche* ^{2}1987: 15; *In Search* 1992: 6)

The idea that it takes a major revolution to unlock the fixation of scientists on one framework, one paradigm, has major consequences for Kuhn's view of the growth of scientific knowledge. As scientists adhere to paradigms as world pictures, which influence them to the point of shaping their perception of the world around them, new theories offer new worldviews. It is the *third* difference with Popper. Knowledge grows by a process of rebuilding or restructuring. It is not only the prior theory, which is reconstructed; prior facts are also re-evaluated. A new paradigm[11] is not just an increment to what is already known. (Kuhn ^{2}1970*:* 7) Growth, according to Kuhn, is discontinuous, in the sense of a complete reconstruction of the edifice of knowledge (Fig. 6.1).

This view of the growth of scientific knowledge, in turn, has major consequences for the conception of rationality. Kuhn agreed that there are criteria for theory choice: simplicity, scope, fruitfulness, predictability, independent testability and empirical success. But the way they are applied in the choice of theories is subject to the

---

[10] Popper insisted on the relationship between his realism and fallibilistic epistemology. The external world is real and independent of human awareness because, in a confrontation of a theory with the external world, the theory may fail. ('Replies' 1974: 1059f)

[11] Kuhn later abandoned the notion of 'paradigm' and replaced it by 'disciplinary matrix', which includes symbolic generalizations, exemplars and metaphysical assumptions. (See Postscript in *Structure* ^{2}1970) His attention also shifted from perceptual to linguistic issues. But these later developments need not concern us here because Popper responded only to Kuhn's theses in *The Structure of Scientific Revolutions*.

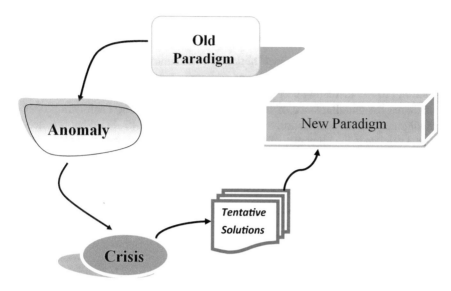

Fig. 6.1   A model of discontinuous growth, interrupted by a crisis and a revolutionary period

evaluation of individual scientists or groups of scientists. They will not decide in the same manner because they adhere to different paradigms. Their preference for theories is influenced by the way they perceive the world. Objectivity 'yes', insofar as it is accepted that different scientists have different views on what counts as objective. Kuhn then placed a strong emphasis on scientific communities and the *tacit* knowledge of its practitioners. The concept of tacit knowledge was as important for Kuhn's as it was for Polanyi's argument. Tacit knowledge is learned by practising science rather than by acquiring rules for doing it. (Kuhn [2]1970: 191) It resembles a skill. Scientific inquiry is a craft. (Popper, however, drew an explicit contrast between objective and subjective knowledge.) Because he regarded scientific inquiry as a craft and scientific theories/paradigms as world pictures, rather than problem-solving tools (as Popper did) Kuhn came to the conclusion that paradigms are closed systems, which hold sway over scientists. This is the *fourth* difference.

The question arises how Kuhn accounted for the possibility of scientific change, if paradigms are such closed systems that they determine the way scientists see the world. Such transitions or revolutions are extraordinary periods in science. They can be understood only against the background of normal science.

Normal science is puzzle-solving on the basis of an accepted paradigm. The researcher is a solver of puzzles not a tester of paradigms. (Kuhn [2]1970: 144) This is a *fifth* difference between Popper and Kuhn. Popper admitted that he had overlooked the existence of 'normal science' but he found the idea 'dangerous'.

> The only criticism which struck me as vital and deserved was – as I have said in my reply to Kuhn - my complete neglect of what Kuhn calls "normal science". I have fully accepted this criticism, though in a spirit which is barely compatible with Kuhn's own. For I see in

the phenomenon called by Kuhn "normal science" a dangerous threat to what I cherish as "science". ('Replies' 1974: 1149; cf. 1145)

What then becomes of the central notion of systematic testing, of attempted falsifications? Systematic testing of paradigms, as Popper recommended, does not happen according to Kuhn. Testing only occurs as part of the competition between two rival paradigms for the allegiance of the scientific community. The falsification of one paradigm equals the verification of its rival. (Kuhn [2]1970: 145–47) What is tested in normal science is the ingenuity of the scientist, not the current theory. If the scientist cannot solve the puzzle this reflects on her/his capacity, not on the theory. S/he, as the problem-solver fails, not the theory. (Kuhn [2]1970: 5)

Popper saw in Kuhn's scheme a threat to science:

...the tradition of problem solving may give way to that of puzzle solving; and the idea of scientific progress (however revolutionary) through problem solving may have to yield to that of the succession of a number of more or less incoherent periods which hardly understand each other. ('Replies' 1974: 1179)

Kuhn (1977: 234) held that the seeds of revolutionary change are built into normal research. A revolution occurs when the profession can no longer ignore anomalies that subvert the existing tradition of scientific practice. (Kuhn: [2]1970: 6)

A crisis precedes the revolution. And the crisis is itself preceded by an awareness of an anomaly. Let's assume that a discrepancy between the theory and the empirical observations turns into an anomaly. That means, according to Kuhn, that some members of the profession accept that the paradigm is in trouble. With the transition from an anomaly to a crisis, extraordinary science sets in. Scientists may try to fix the existing paradigm through minor changes. If the problem persists, then the paradigm itself becomes blurred and the crisis may end with the emergence of a new candidate for a paradigm, with the ensuing battle over its acceptance.

The transition to a new paradigm means that a new tradition of normal research emerges. It is one of the most controversial insights of Kuhn's theory that the emergence of a new tradition is not a cumulative process. It is not achieved through the extension of the old paradigm. Rather Kuhn describes this transition as 'a reconstruction of the field from new fundamentals'. When the transition is complete the field will have changed its view of the paradigm, its methods and goals. The earth became a planet, Kuhn claimed, only after the Copernican revolution. The old and the new paradigm are now incommensurable: they cannot, without residue, be translated into each other. (Kuhn [2]1970: Postscript, p. 202) Scientists who adhere to different traditions experience a partial or total communication breakdown. As Polanyi had already pointed out, logical arguments will be insufficient: scientists must experience a conversion to the new paradigm by means of persuasion. Kuhn admitted, of course, that more than fashion is involved in the determination of scientific progress. A group of specialists must endorse the new paradigm on the basis of accepted values. A new paradigm must fulfil two conditions to be accepted as a new candidate:

1. It must seem to resolve some outstanding and generally recognized problem.
2. It must preserve a relatively large part of the concrete problem-solving ability that has accrued to science through its predecessors. (Kuhn [2]1970: 169)

What emerges in the end is that science grows according to a Darwinian pattern, not towards truth, but towards better and better ways of describing and explaining complex phenomena. What gets selected in the evolutionary struggle is the fittest way to practice future science. (Kuhn [2]1970: 170–3)

As the reader may have guessed, Popper objected strongly to the claims that science does not seek truth, that paradigms are incommensurable and that the growth of science is not a rational affair. Popper also rejected the idea of the monopoly of one dominant paradigm. (*Zukunft* [4]1990: 59; [355.02]) He regarded a paradigm simply as a 'dominant theory'. ('Replies' 1974: 1069) Since *Grundprobleme* and *Logik,* he had spoken of a selection of competing theories. In response to Kuhn, Popper developed his own evolutionary model of scientific growth, rational criteria of its assessment and a notion of approximation to truth (verisimilitude).

6.4 Popper's Response to the Post-rationalists

The starting point is his tetradic scheme of the growth of scientific knowledge, a simple model of which was introduced in the section on falsifiability. I offered it as one of the reasons why testability is a better term for Popper's methodology of science than falsifiability. According to the basic scheme science starts from a problem (P_1), develops tentative solutions (*TS*), eliminates erroneous attempts (*EE*) and ends up with a new, more sophisticated problem (P_2). This simple scheme did not take into account that there will always be competing theories, which try to tackle the problem, and hence different strategies for error elimination. Throughout his work Popper stressed that at any moment in their history scientific theories face competition. In *Logik* he had insisted that 'competing theories can be compared.' (*Logik* 1935/[5]1973: §82 [fn * 2]; cf. §9)[12] A Darwinian selection [*Auslese*] of theories takes place. (*Logik* 1935/[5]1973: §§3, 30; translated into English as 'natural selection') There is competition between rival theories in their attempt to solve a pressing problem. Only the best theories survive as the ones that have withstood severe tests. Hence, Popper never was a naïve falsificationist.

Unlike Kuhn, Popper defended a *cumulative* view of scientific knowledge. (*C&R* 1963: 236, 239) But he did not mean that science grows by piling up facts. Speaking of the 'accumulation method', in a lecture in Oxford in 1948, Popper ruled that it was less important than growth by the method of 'criticism'. 'There is much less accumulation of knowledge in science than there is revolutionary changing of scientific theories.' (*C&R* 1963: 129) On the other hand, continuous growth of scientific knowledge guarantees its rational and empirical character. It is precisely through this kind of growth, via a process of testing (error elimination), that the rational and empirical character of scientific theories is guaranteed. If science does not grow in

---

[12] The English edition translates this phrase as 'that we can *compare* degrees of corroboration' (*Logic* 1959: §82 [fn*2], italics in original). On the elimination of 'unfit theories', see *Objective Knowledge* (1972): 13–7, 264–5.

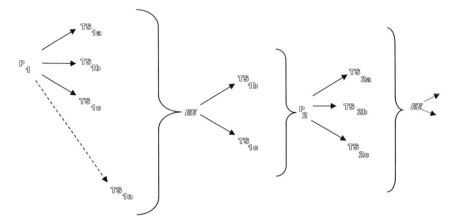

Fig. 6.2 Popper's extended tetradic scheme, reflecting competition amongst tentative solutions

this continuous manner, then it loses its distinctive character. The growth of science is not linked to the accumulation of observational facts but to the development of theories with a higher degree of empirical content or of testability. (*C&R* 1963: 215–17) It is still a cumulative, Lamarckian view of the growth of scientific knowledge because scientific theories seek to approximate truth. The central argument is that theories are 'falsified' and replaced by theories with a higher degree of corroboration. A more realistic picture of this evolutionary process must reflect the presence of alternative and rival theories in a modified tetradic scheme (Fig. 6.2).

6.4.1 Revolution in Permanence

Readers of Popper's later work may have noticed that he referred more often to the *revolutionary* character of science than on previous occasions (for instance in his Introduction to *Realism and the Aim of Science*, 1982). It would be a mistake to understand it in a Kuhnian sense. Progress in science is both revolutionary—established theories are overthrown—but also conservative because it grows cumulatively towards the ultimate aim of truth. (*Myth* 1994: Chap. 1. VIII; *Quantum Theory/Schism* 1982/1992: Preface X) A rational choice is made between competing theories, on the basis of objective criteria. It is noteworthy that in a letter to the cultural historian Egon Friedell—of which several copies exist, the last one dated July 30, 1932 [017.06; 297.22]—Popper called his book *Die beiden Grundprobleme* 'a child of crises, (…) above all of (…) the crisis in physics (…). It affirms the *permanence of crisis*; if this is right then crisis is the normal situation in a highly developed rational science'. (*Grundprobleme*, Foreword to the 1979 German Edition, p. xviii; italics in original; English translation 2009: XVIII) Popper's words are reminiscent of a similar statement from the influential physicist Wolfgang Pauli about the state of physics in the years before 1925:

> At the moment physics is again terribly confused. In any case, it is too difficult for me, and I wish I had been a movie comedian or something of the sort and had never heard of physics. (Quoted in Kuhn [2] 1970: 84)

These words were spoken by a man whom Einstein would nominate for the Nobel Prize in physics, which he received in 1945 for his formulation of the Pauli Exclusion Principle, a fundamental contribution to quantum mechanics.

Popper did not accept that Kuhn's work on the history of science constituted a refutation of his methodology. (*Realism/Aim*: Introduction 1982) If he was influenced by Kuhn, it was in his increasing references to the revolutionary aspect of science. In his exchange with Kuhn and his followers Popper spoke of science undergoing 'revolution in permanence'. ('Replies' 1974: 1147, 1149; see Worrall 1995) He rejected Kuhn's notion of normal science as inimical to the rationality of science. In a letter to the Austrian theoretical physicist, Roman Sexl (31/07/79), he insisted that for Kuhn 'normal science was typical', but 'it is unrevolutionary, even submissive to authority'. Unlike Kuhn, Popper saw 'revolutions everywhere' [349.04; my translation]. What made science revolutionary was the never-ending verdict of critical scrutiny. For Popper 'revolutionary' science is 'critical' science. ('Replies' 1974: 1149) It renders science hypothetical and its results uncertain.

What Popper meant by 'revolution in permanence' is expressed in his evolutionary model of the growth of science. It is captured by his expanded tetradic scheme. This more elaborate view was Popper's final answer to Kuhn's attack. It is a chain of reasoning model because the transitions between theories are reasoned, rational affairs (Fig. 6.3). Science evolves from problems to problems, i.e. problems of increasing depth. (*C&R* 1963: 222) It grows through tentative solutions and error elimination to new, more profound problems. But historians of science will always be able to navigate their way through the maze of tentative solutions, because there are reasoned transitions between, say, Aristotelian physics and Newtonian mechanics. And there are reasoned transitions between Newton's and Einstein's theories. If the velocity of objects approaches the speed of light, c, classical mechanics begins to give way to the Special theory of relativity. If accelerated motion and gravitational effects are taken into account, the Special theory gives way to the General theory of relativity. Popper's scheme resembles what I characterized as inference to the most plausible explanation. In the history of science there are indeed many examples of such inferences. What is matter? Why do planets move? What brings about the diversity of life? All these problems produced a number of tentative solutions and led to the formulation of more refined problems. Recall that in 1651 the Italian astronomer, Giovanni Riccioli, published a textbook of astronomy which presented five different models of planetary motion. In *The Origin of Species*, Darwin contrasted and compared his theory of natural selection with rival accounts, including creationism. He never claimed that his theory was true. It offered a more plausible account, a better fit, of the evidence. Popper himself discussed the history of materialism, a theory which has undergone many different versions. (*Self/Brain* 1977: Chap. P1)

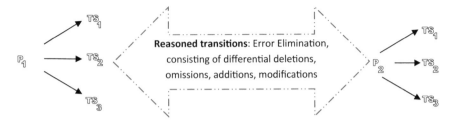

Fig. 6.3 The growth of scientific knowledge as a *chain of reasoning* model. It is possible to follow the evolution in two directions: from the Problem P_1 via tentative solutions TS_n to the problem P_2. And from P_2 back to P_1 via the tentative solutions. According to Popper, scientific revolutions are rational affairs

6.4.2   Corroboration and Testability

Popper asserted that theories are bold conjectures, which aim at a true description and explanation of the natural world. Aiming at a true description does not guarantee that this aim will be achieved. Science undertakes 'serious attempts to discover the truth' but the truth may elude it. Scientists may never be sure that they have hit upon the truth. Popper compared theories in this respect with the status of Goldbach's conjecture: 'every even number (except 2) is the sum of two primes.' (*C&R* 1963: 115, 307) It may or may not be true. Perhaps mathematicians will never know.

Theories, then, are 'highly informative guesses about the world which, although not verifiable, can be submitted to severe critical tests'. (*C&R* 1963: 115) The theories which survive the tests are what Popper calls 'corroborated' theories. The severity of tests increases their degree of corroboration. Scientific theories cannot be justified as true, but scientists can justify their preference for certain theories insofar as it can be shown that they get nearer the truth than their rivals. How can this aim be achieved? If it is found that a certain theory performs better under severe tests than others, then Popperians believe there is good reason to believe that it is getting nearer the truth about the natural world. Popper reserved the term *corroboration* for this process. Corroboration means that a theory has withstood severe tests but this is no guarantee that it will be successful in the future. On the basis of the evidence it is a better approximation to truth than its competitors. It is the fittest in a relative sense: relative to the available evidence. For instance, Ptolemy's planetary theory, which was based on the geocentric worldview, appeared to the Greeks to be well-tested. It seemed to fit the evidence available at the time better than rival theories. For Popper the degree of corroboration of a theory is an *indicator* of its truthfulness at a given time. (*C&R* 1963: 234–35; *Objective Knowledge* 1972: 103) (In a letter to an unnamed correspondent (10/07/35) Popper declared that falsifiability was relative to a 'class of empirical propositions' [017.06]).

Under which conditions can it be claimed that our knowledge of the physical world has been extended? Popper takes a radically negative view. Theories can at best be approximations to truth. No amount of testing can ever make them more probable. If a theory withstands severe tests it is *corroborated*, not confirmed. For Popper confirmation was associated with higher degrees of probability. The better a theory is confirmed the higher its probability. Corroboration, however, means only that a theory has so far survived severe tests. It does not, in Popper's view, involve any claims about the probability of its truth. The degree of corroboration is an evaluating report of past performances: its problem-solving capacity, degree of testability, its ability to pass severity of tests. It makes no predictions about the future performance of the theory. (*Objective Knowledge* 1972: 17–21)

It is natural to wonder how scientists can know that they are getting nearer the truth, if the truth may forever elude them. Popperians answer that there is no need to know. Now even critical rationalists may hesitate to accept this point. Ptolemy's geocentrism gave the Greeks a good reason to prefer it over Aristotle's homocentric model but it was not in any obvious sense closer to the truth.[13] Popper relied on an 'old argument', namely.

> …that concatenations of accidents are very improbable; so that if a theory passes many varied tests it is highly improbable that this is due to an accident; highly improbable therefore that the theory is miles from the truth. ('Replies' 1974: 1041, 1011; cf. 'Two Faces of Commonsense' in: *Objective Knowledge* 1972: 101–103)

A determined critic may still find counterexamples to this 'old argument'. But as a pragmatic criterion it certainly makes sense. The idea that a theory T_2 turns out to be better corroborated than a theory T_1 fits in well with Popper's basic view that science is aiming at the truth (= correspondence to facts) but that scientists should remain sceptical about the future success of their theories. They may be corroborated because they have passed severe tests. But they are not 'confirmed' as they may eventually be replaced by better theories (with better approximation to the truth). It is tempting to conclude that a higher degree of corroboration implies a higher degree of verisimilitude; a conclusion which Popper denies. The degree of corroboration of a theory is a measure of the severity of tests it has passed. It is not a measure of verisimilitude. (*Objective Knowledge* 1972: 103)

6.4.3 Corroboration and Verisimilitude

If Popperians wish to hold that one theory gets closer to the truth than any of its competitors, they need some criteria on which people can agree. Popper proposed a number of such criteria of how one theory, T_2, can claim greater approximation to truth than its predecessor, T_1.

i. T_2 makes more precise assertions than T_1, and they stand up to more precise tests.

---

[13] According to Aristotle's homocentric model, the Earth is located at the centre and concentric shells (orbs) carry the planets around it. The outer ring carries the distant stars. The model fails because planets move at varying distances from the Earth, as the Greeks knew from observations.

ii. T_2 takes account of, and explains, more facts than T_1 (including i).

iii. T_2 describes and explains the facts in more detail than T_1.

iv. T_2 has passed tests, which T_1 has failed.

v. T_2 suggests new experimental tests, not considered before (perhaps not even applicable to T_1) and T_2 passes these new tests.

vi. T_2 has unified and connected various hitherto unrelated problems. (*C&R* 1963: 232)

The General theory of relativity predicted the bending of light near gravitational bodies, which was corroborated by Eddington's expedition (Fig. 2.1a). Newtonian mechanics unified two hitherto unrelated phenomena: falling apples and the orbits of planets under the concept of gravitation. The evolution of species and their geological distribution was unified under the concept of natural selection. In particular, a theory must identify the parameters between which it establishes relations (for instance, mass and gravitational attraction). Both the parameters and the relations must be quantifiable. Such quantifiable relations between parameters often form the basis of the laws of science. Astrology stumbles over such a requirement. The positions of the stars in the sky can be stated with numerical precision but the same is not true of character traits. This drawback makes it difficult to quantify any supposed correlation between character traits and the position of the stars.

These criteria can be summarized under the notion of *testability*. Testability is a *contrastive* activity. (See Sober 1999: §1) After all, Popper's criteria are based on a contrast between T_1 and T_2. The easiest situation occurs when T_1 makes a prediction, Pr_1, whilst T_2 makes a different prediction, Pr_2. If, on a simplified Popperian scheme, Pr_2 does not occur, T_2 is regarded as falsified. If Pr_1 is observed, T_1 is corroborated. What happens, however, if T_1 and T_2 make the *same* prediction, say, regarding the motion of Mars, and T_1 and T_2 are based on different principles, respectively? Then further tests will be required until the two theories can be clearly distinguished. That is, central elements of theories may have to be tested.

The suggestion that theories make *bold* conjectures implies that they become *less* probable. The more a theory says, the more it exposes itself to the risk of being refuted. A theory is the bolder, the greater its empirical content. It is also the riskier theory. (*Objective Knowledge* 1972: 53) The testability of a theory increases and decreases with its informative content and therefore with its improbability. (*Objective Knowledge* 1972: 17–21) As Popper pointed out scientists are not interested in trivialities but in solutions of deep problems. The idea of greater empirical content implying greater testability is the logical basis of the method of conjectures and attempted refutations.

Adherence to Popper's criteria means that the empirical content of a theory T_2 will be greater than that of T_1. The empirical content gives rise to the notion of verisimilitude. (Box 3) A theory's verisimilitude, its truthlikeness, is greater if it has greater truth content (but lesser falsity content) than its rival. Popper acknowledged that he was not the first to speak of the growth of scientific knowledge but

> ...nobody, it seems, suggested that the growth of knowledge should be identified with the growth of the empirical content of our theories. ('Replies' 1974: 1036)

The empirical content of a theory comprises 'the class of all basic statements which contradict it, i.e. the class of potential falsifiers'. (*C&R* 1963: 385; cf. Simkin 1983: Chap. 8) (Box 3)

Box 3: Popper's Notion of Verisimitude

Popper defines *verisimilitude* by reference to the notion of content. The class of all the true statements, which follow from a given statement (or which belong to a given deductive system) and which are not tautological is called its *truth content*.

The class of false statements entailed by a statement—the subclass of its content which consists of exactly all those statements that are false—is called its *falsity content*. (*Objective Knowledge* 1972: 48)

The verisimilitude of a statement increases with its truth content and decreases with its falsity content. The problem of verisimilitude arises from the question how one false theory can be closer to the truth than another.

A theory t_1 has less verisimilitude than a theory t_2 if and only if (a) their truth contents and falsity contents (or their measures) are comparable, and either (b) the truth content, but not the falsity content of t_1 is smaller than that of t_2 or else (c) the truth content of t_1 is not greater than that of t_2 but its falsity content is greater.

In brief, t_2 is nearer the truth, or more similar to the truth, than t_1, if and only if more true than false statements follow from it, or at least equally many true statements but fewer false statements. It then corresponds better to the facts.

Although Popper called this idea the 'intuitive comparability of the contents of two theories, t_1 and t_2' (*Objective Knowledge* 1972: 52), his efforts were to no avail. Several researchers, including David Miller, realized that no two false theories can be compared with respect to verisimilitude. In other words, no false theory is closer to the truth than any other. Many letters between Miller and Popper were devoted to this problem. Miller's *LSE* M.Phil./Ph.D. thesis (1974) was entitled 'Some difficulties in the theory of verisimilitude' [328.02].

Popper more or less abandoned his attempt at precise formulations of the notion of verisimilitude. In response to Ayer he accepted that the notion was formally deficient but still not useless. It was not a criterion to assess progress towards truth. ('Replies' 1974: 1100ff and [fn165a]; see also his reply to Suppes, *ibidem*: 1125ff) Nevertheless, work on the notion of verisimilitude continues. (Miller 1994; 2014)

Above I have stated Popper's criteria for a choice between two competing theories, T_1 and T_2. On a higher, evolutionary level, the growth of scientific knowledge should adhere to the criteria of simplicity, independent testability and empirical success.

- The new theory should proceed from some simple, new and powerful unifying idea about some connection or relation (like gravitational attraction) between hitherto unconnected things (such as planets and apples) or facts (such as inertial and gravitational mass) or new theoretical entities (such as fields and particles). This criterion is called *simplicity*.

- The new theory should be independently testable. Apart from explaining all the explicanda which the old theory was designed to explain it must have new and testable consequences. It must lead to new predictions, which have so far not been observed. This criterion is designed to exclude ad hoc hypotheses. (A theory is ad hoc if the only 'evidence for the explicans is the explicandum itself'.) (*Objective Knowledge* 1972: 192)[14]

The first two criteria provide potential progress, whatever the outcome of the new tests. The new theory explains all the explicanda of the previous theory, and in addition it gives rise to new tests. But it is not sufficient for a new theory to (a) present a unifying idea about the structural properties of the world (*simplicity*) and (b) lead to predictions of new unobserved phenomena. These two points are formal requirements. The new predictions must be tested, which gives rise to the third requirement of scientific growth.

- The new theory should pass some new, and severe tests, i.e. it must achieve some empirical success. It should not be abandoned too soon. This is a material requirement. A new promising theory should not be discarded simply if it fails, for the time being, to produce startling results. Scientists sometimes have confidence in a theory even though they know that it has deficiencies. (*C&R* 1963: Chap. 10, §5; *Objective Knowledge* 1972: 30)

Einstein supposedly claimed that if Eddington's expedition (1919) had failed to show the deflection of starlight, bad luck, the theory (of general relativity) was still right. Popper accepted that a limited amount of dogmatism, pigheadedness, was necessary for progress. (*Myth* 1994: 16; *Realism/Aim* 1983/1985: 70) However, it should not be forgotten that the degree of uncertainty is a function of how firmly scientific theories are embedded in the web of knowledge. There is clearly a difference between speculations about the multiverse or the possibility of time travel and the reliability of Newton's laws of motion. At the coalface of scientific research there is more tentative groping towards the truth than in established areas. Some areas of science enjoy more reliability than others. Popper himself conceded that some conjectures were 'better' than others. His *principle of preference* stated that 'a theory with great informative content is on the whole more interesting, even before it has been tested, than a theory with little content.' ('Replies' 1974: 1022) In *Grundprobleme* (1979: 96, 2009: 86) he admitted that there is a gradation in the certainty of hypotheses.

---

[14] An example of what looked at first like an ad hoc hypothesis, but which turned out to be independently testable, was Darwin's appeal to the poverty of geological records. Darwin was trying to deal with the objection that there should be missing links between past and present species. If natural selection is right there should be 'an indeterminable number of intermediate forms'. (Darwin 1859: 462) Such geological records of intermediate forms were eventually found.

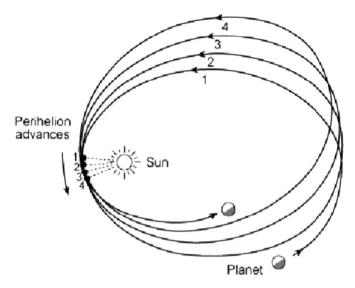

Fig. 6.4   An Illustration of the Perihelion Advance of Mercury

Theories replace each other on the ground of better corroboration. In this process we also learn more and more about the physical world. Scientific theories extend their scope and explanatory depth. There are clear examples of the growth of scientific knowledge. For instance, according to Newton's theory Mercury's perihelion point (its closest point to the sun) advances by 532″ (arcseconds) per century, due to the perturbations of other planets and the sun. But this left an observational discrepancy of 43″ (arcseconds) per century, which Einstein's General theory of relativity was able to explain (Fig. 6.4). The perihelion precession of Mercury was a problem in the Newtonian theory because it had not been able to account for this observed difference of 16° in 10,000 years.

Popper's emphasis on the logical evaluation of competing theories had consequences for how he assessed the historical understanding of problem situations, which constituted a further difference with Kuhn.

6.5   On Historical Understanding

Popper's criteria demonstrate that he was a fallibilist, not a naïve falsificationist. (*Grundprobleme* 1978: §1; 'Replies' 1974: 1054) The growth of scientific knowledge can only be labelled 'rational' if scientists systematically choose the better theory in any problem situation, according to Popper's requirements of growth.

Popper also disagreed with Kuhn on the question of how the history of science should be written. For Kuhn the task of the historian of science was the description of the actual historical situation in its integrity. The historian had to assess a particular scientific period on its own terms, including criteria of rationality and reality. Kuhn demonstrated this method of historical integrity—the refusal to impose today's criteria on yesterday's paradigms—in his masterful analysis of the history of astronomy from the Greeks to the Newtonian revolution. Popper took a different view. Looking at the history of science, the main aim of all historical understanding was the hypothetical reconstruction of a historical problem situation. It was the application of *situational analysis:* a logical assessment of the solution a historical agent envisaged when grappling with a certain problem. Given that historical agents faced a given problem—a Copernicus, Newton or Darwin—were their solutions adequate to the problem as they perceived it? Popper called the method of *situational analysis* the application of the rationality principle. (*Objective Knowledge* 1972: 179, 170ff, Chap. 4) Whilst in his earlier work it referred to models of social agents, in the present context it refers to an objective reconstruction of the historical problem situation. Popper preferred to speak of the objectivity of knowledge without invoking the producers of knowledge. As he put it dramatically: objective knowledge 'is *knowledge without a knower'*. (*Objective Knowledge* 1972: 109; italics in original) There was no room for Kuhn's *gestalt* shifts or Polanyi's conversions. To uphold his views on objectivity Popper developed a distinction between subjective and objective knowledge. In this way he arrived at his theory of the three worlds. Subjective and objective knowledge belong to different worlds, which in turn differ from the material world.

6.6   On the Objectivity of Knowledge

As he freely admitted Popper was not the first to locate objective knowledge in an 'immaterial' realm. (*Autobiography* 1974: Chap. 38; *Knowledge* 1994: Chap. 3) He followed Plato, the mathematician G. Frege and the philosopher and logician B. Bolzano in locating objective knowledge – theories, proofs, problems – in a 'third world', although his characterization departed from that of his predecessors. Under the influence of John Eccles, who had become a Popperian, Popper renamed it World 3. He had not used that terminology before 1966. (*Autobiography* 1974: 157 [fn7a]) World 3 contains the logical contents of books, libraries, computer memories, journals. (*Objective Knowledge* 1972: 74) Scientific knowledge also belongs to World 3, that is, the independent world of theories, problem situations as well as critical arguments. Knowledge in this objective sense is said to be independent of anyone's claim to know or to believe. It is knowledge without a knowing subject. Like Frege, who used the term 'third realm', Popper bestowed autonomy on World 3. This implies that World 3 harbours unknown territory, which scientists can explore.

Popper's handwritten note about Worlds 1, 2, 3 for a conference in Turin (12–16.01.1983). *Source* Wikimedia Commons

Although it was more an epistemological than an ontological category, it led to misunderstandings. Feigl, for instance, felt uncomfortable with its 'ontological flavour' (letters to Popper 15/01/72 [294.06] and 31/12/71, 21/11/74 [536.10]). One of Popper's followers, Günter Wächtershäuser—a German chemist who later became a patent lawyer—was interested in the 'three-world scheme' because it could be used to solve 'practical problems'. In particular, as he explained to Popper (11/11/83), the theory serves to tackle problems in patent law, i.e. 'the evaluation of the merits of an invention'.

It has been found that for such evaluations, it is necessary to study the problem solved by an invention. In the early days of patent law, the subjective problems of the inventor were studied. Later it was found that such problems are rather insignificant and that it is necessary to determine the objective problem of an invention. (…) It is quite apparent that the court decisions on the "problem of an invention" refer to problems as world-3 entities rather than world-2 entities [359.02, underlined in original].

In addition to World 3, Popper distinguished World 1—the physical world—and World 2—the world of our conscious experiences and awareness. Knowledge in the subjective sense, World 2, consists of mental states: beliefs, subjective feelings, behavioural dispositions.

There are feedback loops between World 3 and World 1. When Francis Bacon proposed that knowledge is power, he meant that the dominion of nature was to be achieved via knowledge about the physical world. Such knowledge allows scientists to manipulate the material world to a certain extent. Natural energy is harnessed through the application of scientific knowledge. It may, however, be found that physical objects in World 1 do not agree with World 3 theories about them. It is then a case of recalcitrance, which has always forced scientists to change their World 3 knowledge claims. The independence of World 1 guarantees the testability of scientific theories.

The world of mental states, World 2, and World 3 also interact. Popper followed Kant in asserting that the mind imposes order on nature. A conscious mind has to use symbolic language to formulate a scientific law before it becomes public knowledge. An intelligent being must discover the laws of science through a mental process. In this sense World 2 makes a contribution to World 3. By making it public the discoverer or inventor deposits the discovery in World 3. The mental product—be it a scientific theory, law or an equation—enters World 3, as an objective proposition, or a new problem. It can now be discussed independently of the mental processes through which it came to inhabit World 3. Conversely, World 3 problems lead to creative thinking in World 2. Intelligent agents choose a World 3 proposition and contemplate it in the privacy of their minds. Popper held that subjective knowledge, World 2 knowledge, depended upon World 3 knowledge, i.e. on linguistically formulated theories. (*Objective Knowledge* 1972: 74) The language in which such theories are cast is public. A human language is a public property before any individual makes use of it.

Is there also an interaction between World 1 and World 2? This is the notorious mind–body problem, to which Popper and Eccles offered a contribution in the 1970s. Their book, as we shall see, defends an interactionist theory of the mind. The distinction between the three worlds prepared the way for Popper's interactionism in the philosophy of mind.

Despite the interaction with other worlds, World 3 must be considered to be largely autonomous. It consists of publicly available knowledge, ready to be picked up when needed. It also survives generations. It is not dependent for its survival on whether a particular mind is preoccupied with one of its items at any particular time.

Eccles proposed two thought experiments to illustrate the 'independent existence of World 3':

> In experiment (1) all machines and tools are destroyed, also all memories of science and technology including our subjective knowledge of machines and tools, and how to use them. But *libraries and our capacity to learn from them* survive. (…)
>
> In experiment (2) the destruction is greater, because not only is there the destruction of World 1, but in addition *all libraries are destroyed*, so that our capacity to learn from books becomes useless.

Under the first scenario, Eccles concluded, humankind would eventually be able to restore civilization because the knowledge to do so was available in World 3. The second thought experiment leads to a much bleaker conclusion because humanity would plunge into barbarism and would have to relearn everything that had been lost. (Eccles 1974: 351; italics in original)

Yet how does knowledge in World 3 grow? It can only grow if an intelligent agent, equipped with symbolic language, interacts with it. This language must contain the argumentative function. Only an agent who has the mental capacities of dealing with the problems stored in World 3 can work on them, modify them and change them. Humans, as conscious beings, interact with World 3 and hand the results of their mental operations and creations back to World 3. In this way Popper arrived at his evolutionary picture of how our knowledge is supposed to grow. World 3 contains many competing theories, leading to many error eliminations, tentative solutions and new problems. (*Objective Knowledge* 1972: 287) (Fig. 6.5)

Continuous growth, rather than constant remodelling, assures us of the rationality of science. The 'growth of our scientific knowledge is the result of a process closely resembling "natural selection"'. Only the fittest survive. It is not the natural selection of species but of hypotheses or tentative theories and solutions.

> Our knowledge consists, at every moment, of those hypotheses which have shown their (comparative) fitness by surviving so far in their struggle for existence; a competitive struggle which eliminates those hypotheses which are unfit. (*Objective Knowledge* 1972:261)

For Popper this is more than a metaphor. It is a description of how knowledge grows:

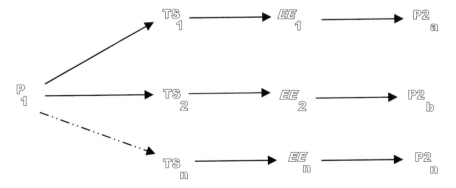

Fig. 6.5 Towards ever more sophisticated problems, through 'trial and error'. Each tentative solution is subject to error eliminations

> From the amoeba to Einstein, the growth of knowledge is always the same: we try to solve
> our problems, and to obtain, by a process of elimination, something approaching adequacy
> in our tentative solutions. (*Objective Knowledge* 1972: 261; cf. Munz 1993)

This evolutionary process leads to greater specialization and differentiation. New disciplines, like computer science, molecular biology or particle physics branch out of the common trunk. Rapidly, people who specialize in one of those disciplines become fluent in the currency of new theories. But Popper was critical of intellectual fashions and specializations. They are part of the belief in the authority of the experts. He warns us that

> ...the orthodoxy produced by intellectual fashions, specialization, and the appeal to author-
> ities is the death of knowledge, and (...) the growth of knowledge depends entirely upon
> disagreement. (*Myth* 1994: x; cf. *Realism/Aim* 1983/1985: 259f)

From this shift to evolutionary thinking it was not a giant step to Popper's reflections on evolution and the status of evolutionary theory.

In the early part of his career Popper was mostly influenced by Newton and Einstein. Later he turned to Darwin's evolutionary thinking. He developed a dynamic view of time. Was he also indebted to his fellow philosophers? He had numerous discussions with members of the *Vienna Circle* but he disagreed with them on many points.

Popper experienced some ups and downs in his relationship with other philosophers. I have already discussed his quarrels with Lakatos and Agassi. Further disputes erupted with Bartley and Watkins. He did not think highly of most of his fellow philosophers. Abiding influences, however, were Kant and Russell. And the Greeks.

Chapter 7
Popper and His Fellow Philosophers

We don't need thousands of trained and badly trained philosophers – it is very silly. Actually most of them have nothing to say. (K. Popper, quoted in the Lesley White Interview, *The Sunday Times*, News Review 12/07/1992, p. 2)

As many commentators have observed Popper was hungry for recognition. This may be unsurprising, given the stuttering start to his academic career. A genuine modesty veiled his need for recognition. He had withdrawn to the Chiltern Hills to live a simple life. The only extravagance was a large piano in his living room at *Fallowfield*, in Penn. He complained in an interview, falsely as we have seen, that his contributions to quantum theory had not been acknowledged. As it turned out, his views on evolutionary biology were ignored by many biologists. Popper was not well-liked by some of his fellow philosophers. Contemporaries, like Schlick, considered him difficult. 'He was brilliant, but self-focused, both insecure and arrogant, irascible and self-righteous'. (Hacohen 2000: 209, 318, 321)[1] The impressions of philosophers' spouses were also divided. Research on Popper's biography led Bartley to the Carnap Archive in Pittsburgh, where he came across an exchange between Mrs. Carnap and Mrs. Hempel (1936). He openly told Popper (29/03/83) that Mrs. Carnap felt hostile, and Mrs. Hempel felt friendly towards him [273.03]. Popper returned the compliment by a general dislike of his philosophical colleagues, with which Schilpp agreed (letter dated 13/08/69, [334.02]). Popper revealed to Peter Munz (14/03/86) that he found all works written about him 'irritating', with the exception of Magee's book. So he would not read Munz's planned book on Popper and Wittgenstein. He chided Munz for his careless reading of his work and added that 'what you say about truth is bad'. Munz's response was apologetic (10/09/87) [553.10]. His fellow philosophers either had 'nothing interesting to say' or tended to misunderstand him.

---

[1] Hacohen (2000: 209) reports that already during his Vienna years, members of the *Vienna Circle*, such as R. Carnap and V. Kraft, found him brilliant but difficult to deal with.

© Springer Nature Switzerland AG 2022
F. Weinert, *Karl Popper*, Springer Biographies,
https://doi.org/10.1007/978-3-031-15424-9_7

7.1 Bedford College Conference

Of particular significance was a conference held at Bedford College in London on July 11–17, 1965. It occurred on the heels of Kuhn's monograph on the importance of scientific revolutions in the history of science. Some of the papers from this colloquium were later edited by I. Lakatos and A. Musgrave under the title *Criticism and the Growth of Knowledge* (1970). Lakatos's paper was not presented at the colloquium. This meeting of influential philosophers of science was important for personal reasons, too: it was a turning point in Popper's relationship with William Bartley. The two men had known each other since 1958, when Bartley became Popper's graduate student. Popper described him as 'the most outstanding young philosopher I have met' [271.25]. They quickly became close friends. Popper invited Bartley to Penn for Christmas in 1960 and acted as his referee. But in his contribution to the Bedford College conference Bartley attacked Popper's falsifiability criterion of demarcation as a blunt weapon to distinguish scientific from non-scientific theories [272.03]. Lakatos, in his role as Colloquium Secretary, asked Bartley to withdraw the paper (13/07) because he felt it was disrespectful to Sir Karl. But Bartley went ahead and delivered his talk. Popper declined to be drawn into this spat (as Watkins had advised him [359.34]). Bartley did not let things rest. He addressed an open letter to the President of the British Society for the Philosophy of Science, Professor Stephan Körner of the University of Bristol, in which he accused Lakatos of censorship because he had threatened to stop the publication of his talk. In the following weeks Lakatos withdrew his threat and expressed his wish to include the paper in the Proceedings. But Bartley was so incensed that he refused to have his paper published, as long as Lakatos was the editor. This whole affair prompted the Society to issue a report (09/07/66), which did not uphold Bartley's accusations. The disagreement dragged on until the end of 1966 when Bartley relented (13/12/66). He had been informed that the Proceedings would be edited by Musgrave and Lakatos, which prompted him to give his permission to publish his contribution, not in the 1970 volume but an earlier collection (Bartley 1968) [272.03]. Popper was profoundly shocked by Bartley's attack. (His criticism was harsh but apart from typical statements such as 'Popper is mistaken' and 'Popper is wrong', there is nothing in the way of a personal attack. Yet Popper took the criticism personally.) He wrote a long reply, in which he expressed his dismay [271.22]. He also sent a personal letter to Bartley (20/07/65), in which he explained that he found his talk 'incomprehensible' but was 'willing to look upon it as if it had never happened…let us be friends again', he added. 'Life is short, and one has not so many friends to throw away' [272.02]. Their estrangement lasted until 1974 when Bartley recontacted Popper to congratulate him on the publication of the Schilpp volume. Popper was overjoyed to receive Bartley's letter [272.05]. He had been deeply unhappy about their split, as he admitted to Bartley on November 11, 1978 [272.07]. In an unpublished interview with David Miller (17/12/92), Popper returned to this unhappy episode. He blamed himself for the rift. Bartley had criticized certain passages on rationality in *Open Society* (especially Chapter 24). Popper accepted this criticism but asked him to revise it. Bartley did, but Popper feared that his friend had been hurt by his apparent failure to acknowledge adequately Bartley's contribution. He speculated that Bartley's perception may have led to his attack at the colloquium. More than a decade later, Bartley revealed to Popper (23/10/78) that he had been annoyed that generally no one had properly acknowledged his 'work on rationality' [272.07]. This perception may have been unfounded because Popper does express his gratitude to Bartley in both volumes of *The Open Society* [552.03].

John Worrall, who was appointed as a lecturer at the *LSE* in 1970, visited him at Kenley (136, Welcomes Road, Croydon, CR8 5HH) south of London where he had moved after Hennie's death (1985). He recounts a revealing incident:

Either late 80s or early 90s (I would have to check), when we were just setting up the Centre and wanted a big publicity talk. He first said he would not see me because of some intellectual crime I am supposed to have committed. And then he let it be known that he would so we arranged to go out there. First he was the real European gentleman, offering me coffee and biscuits and we were chatting nicely about intellectual things, nothing very deep; and then all of a sudden he started to go red from the neck upwards and started to tremble and he launched into an attack on me. I'd never been to see him before was apparently a big crime. I was a nice working class lad; I would have loved to have gone to see him before but I was waiting for an invitation, rather than invite myself. That was one thing. Then again that alleged failure to quote him in a paper that I had never written. And then a long diatribe against Lakatos and how I was Lakatos's protégé and therefore I was guilty by association. He obviously did not say that but that was the sort of input.[2] I was trying to placate him, very sorry but it certainly was not deliberate, I have enormous respect for you and your work.

Then Popper calmed down.

Anyway, then after ten minutes or so of this diatribe, it was really amazing, the blood drained back and he stopped shaking and said: "Come let us talk of more congenial things." And he started telling me about some ideas he'd had about Newton and his scientific method. I mean I was already pretty much of an expert on Newton and it was not accurate but still it was

---

[2] Worrall may have forgotten an article he wrote in June 1982 for the *LSE* Newsletter [273.02]. The article discussed new directions of research and a shift to the history of science. Popper must have been aware of this piece, since Bartley mentioned it in a letter to him (06/07/82). Bartley called the article 'unbelievably bad' [273.02]. In the article Worrall emphasized Popper's legacy of clear thinking and the 'crucial importance of first understanding a thinker's problem situation'. But then he highlighted the problem of verisimilitude and proceeded to praise Lakatos's research programme, which 'inspired a series of historical case studies'. He also concurred with Lakatos's view in general. A one-day Popper conference at the *LSE* (30/04/88) featured Worrall's talk on 'Popper's failure to solve the problem of induction' [552.02]. Popper could not attend but sent Worrall some excerpts of his work on 'inductive probability', which he complained, seemed to be 'quite unknown in my old Department' (22/04/88) [552.03].

interesting and we talked about music a bit. By that stage we were in the room with the grand piano and I admired his library. He had this wonderful, wonderful….he was a great collector of books. And we parted and he said he would think about it, think about the invitation and a couple of days later I heard that he did not want to do it. That was it, that was the first and last time that I visited him really. It was a very notable occasion. (Interview with John Worrall 24/10/2018)

Popper's criticism of his professional colleagues and intellectuals generally would become more severe. He accused them, as he had charged Plato, Hegel and Marx, of an abrogation of their duties towards society, even of intellectual crimes against humanity. Their crimes consisted in having yielded to the temptation of relativism, subjectivism and irrationality. (*Auf der Suche* [2]1987/*In Search* 1992: I.1, III.13; *Alles Leben* 1996: 247, 251–2; *Lesson* 1997: 86, 90) There were exceptions. Feigl, for instance, joined him in his fight against subjectivism and obscurantism' (letters to Popper 15/12/69 and 16/08/72 [294.06]).

There were other notable exceptions. Amongst the Greeks he expressed great admiration for Socrates. Later he devoted much of his time to a study of Parmenides who lived in the fifth-century BC. He had great esteem for the British philosopher and mathematician Bertrand Russell. But no philosopher commanded greater respect and admiration than Immanuel Kant, whose influence over Popper's thinking runs through his whole oeuvre, starting with *Grundprobleme*.

His admiration for Socrates was due to the Socratic method of asking questions and evaluating answers in order to encourage critical thinking. Popper praised Socrates for his frank admission of ignorance and his rationalism. (*Auf der Suche* [2]1987: I, p. 43; *In Search* 1992: 33; *Open Society* I: Chap. 10, §V) Socrates embodied the spirit of scientific enquiry: he was self-critical, committed to individualism and a humble researcher of truth. Socrates was also eager to learn from his mistakes. In his *Autobiography* (1974: §1) Popper thanked his old master, cabinet maker Adalbert Pösch, for having turned him into a disciple of Socrates. Other Greek philosophers whom Popper mentioned with admiration are Xenophanes and Parmenides. He praised both Parmenides and Xenophanes as the initiators of the 'sceptical tradition'. Xenophanes was a man who anticipated the European Enlightenment.[3] Parmenides was important for Popper because the study of his work helped him develop a more sophisticated notion of realism than his often-professed common-sense realism. It will be introduced below as a form of 'structural realism' (Chap. 8). Parmenides drew a distinction between appearances and the underlying reality, which later also appeared in Kant's writings. But Parmenides taught that the underlying reality was unchanging and invariant—a view that Popper sought to refute and to replace by a dynamic view of time and reality.

---

[3] Popper makes a number of references to Xenophanes: *C&R* 1963: Chap. 5; *Auf der Suche* 1985: Chap. 14, §3; *Parmenides* 1998: Essay 2, Essay 6, §6; Addendum.

Bertrand Russell (1872–1970), supporter of Popper. *Source* Wikimedia Commons

7.2   Bertrand Russell

Popper admired Bertrand Russell as a 'man and philosopher'. (*C&R* 1963: 365) Russell quotes are scattered throughout Popper's work. They had met in 1936 when Russell lectured on 'The Limits of Empiricism' to the Aristotelian Society. Popper would later serve as President of that body (1958–59). Their paths crossed again in Cambridge in 1946 when Popper gave his talk on philosophical problems in the presence of Wittgenstein. Russell wrote a letter of recommendation to Simon and Schuster in America in an effort to find a US publisher for *Open Society*. Popper prepared a review of Russell's *History of Philosophy* (1946) for the Austrian Broadcasting Service (1947). 'What makes the book great', he argued, 'is the man who has written it'. [345.14] He admired his courage to change his mind. He expressed admiration for the clarity of his thinking. He compared him favourably to Descartes, Locke, Hume and Kant. But he contrasted Russell with the trendy philosophers of his own time. Russell had saved philosophy from lapsing into 'the intolerable fashion of our time, and into charlatanry and wind-baggery' [345.14].[4] In a letter of May 29, 1959, he asked Russell whether he would accept the following dedication to Popper's *Postscript*:

> To Bertrand Russell whose lucidity, sense of proportion and devotion to truth have set us an unattainable standard of philosophical writing [345.14].

Russell was pleased to accept the dedication (June 9). Things turned out otherwise for the *Postscript* was not published until the beginning of the 1980s.[5]

---

[4] See Grattan-Guinness (1992: 3–18) and (1998: 25–42); both articles appeared in *the Journal of the Bertrand Russell Archives*.

[5] Volume 1 (*Realism and the Aim of Science*) is an extended discussion of inductivism; it is dedicated to William Bartley; Volume 2 (*The Open Universe*) is a defence of indeterminism; it is dedicated to Ernst Gombrich; and Volume 3 (*Quantum Theory and the Schism in Physics*) deals with quantum

Russell was the co-author (with A. Whitehead) of the seminal *Principia Mathematica* (1910–13), which is an investigation of the foundation of mathematics. Russell, the philosopher, had asked the fundamental question: 'How do I know what I know?' (*Realism/Aim* 1983/1985: 86–7, 92) This question had exercised Popper's mind in his debate with the philosophers of the *Vienna School*. He saw Russell as their 'spiritual father'. Russell had also taken the problem of induction seriously. He felt the force of Hume's problem, although Popper did not share Russell's 'subjectivist epistemology', which was based on the empiricist notion of 'sense data'. (*Realism/Aim* 1983/1985: 53–4) Nor did Popper agree with Russell that induction had to be taken as a fundamental a priori principle of science.

Russell had 'discovered' Wittgenstein, whose *Tractatus* had greatly impressed him. But he shared Popper's strong antipathy to ordinary language philosophy or the Oxford School [345.14]. Nevertheless Russell himself had made a significant contribution to analytic philosophy with his influential paper 'On Denoting' (1905). In particular Russell provided a logical analysis of ordinary language expressions. Logical analysis shows how ordinary language can mislead. People say 'All people are not liars' when they mean to say that 'Not all people are liars'. A logical analysis of the first sentence shows that it really means 'Nobody is a liar' (logically: 'Of all people it can be affirmed that they are not liars'). But that is not what people want to express. They really want to say, logically speaking, that 'there are at least some people who are not liars'. Russell also investigated the distinction between meaningful and meaningless statements. (*Open Society* II: 293 [fn 46]) For instance, why is Chomsky's famous example 'Green ideas sleep furiously' a meaningless sentence? Popper, too, was interested in logical analysis. (*Logik* 1935/⁵1973: §§21, 23) He conceived falsifiability as a logical relation between theory and basic statements. Later he developed his situational analysis. He was, like Russell, interested in truth and a defender of rationality.

Immanuel Kant (1724–1804). *Source* Wikimedia Commons

---

theory and presents his propensity theory of probability; it is dedicated to John Eccles and Peter Medawar.

7.3 Immanuel Kant

Still Popper's greatest admiration was reserved for Immanuel Kant. Unlike Popper, Kant never left his home town of Königsberg, which was then part of Prussia. Kant's towering influence is felt throughout Popper's work. By his own admission he would have remained a lifelong Kantian, had it not been for the Einsteinian revolution, as encapsulated in the General theory of relativity. Einstein's revolution in physics, in Popper's eyes, showed that well-established and successful theories, such as Newton's mechanics (with its law of gravity), could be overthrown. Kant's belief that Newtonian physics was necessarily true had been shown to be mistaken. All knowledge, Popper concluded, was conjectural. But did the 'refutation' of Newtonian mechanics not also throw serious doubts on Kant's philosophy? There are a number of aspects of Kant's critical philosophy, to which Popper remained committed. He became a modified Kantian. He characterized his own Critical Rationalism as the crowning of Kant's critical philosophy. It put the 'finishing touches' to Kant's philosophy. (*C&R* 1963: 26–27) A critical attitude towards authority, dogma and tradition was a cornerstone of the Enlightenment period, of which Kant's work constitutes the culmination. Kant famously defined the Enlightenment as 'man's emergence from his self-imposed immaturity'. He saw this immaturity as the inability to think for oneself. Hence his motto of the Enlightenment is *sapere aude*, that is, to have the courage to use one's own reason. It was a rallying cry which echoes throughout Popper's work. As discussed above Popper used Kant's motto to distance himself from Berlin's negative characterization of the notion of positive freedom. Popper, as the 'last laggard of the Enlightenment', extended the range of Kant's critical philosophy to modern science.

There are many aspects of Kant's work that attracted Popper. He admired Kant's early work on the 'island universe', which came to be known as the Laplacean-Kantian nebular hypothesis. In his *Theory of the Heavens* (1755) Kant proposed that what appears to be stars in the sky are actually galaxies. Furthermore, they undergo an evolution such that the formation of cosmic order would take millions, not just thousands of years. Kant held that

- the whole order of nature is still unfolding,
- that law-like regularities extended from the solar system to the Milky Way and the whole Cosmos,
- the whole Cosmos existed of ordered systems, i.e. solar systems, galaxies and clusters of galaxies,
- cosmic order was the result of the operation of *mechanical* laws,
- the gradual establishment of order out of chaos required vast amounts of time,
- Kant's cosmology undermined the chronology of the Bible: it required *millions*, not just thousands of years for the unfolding of the universe.

What is more, Kant, too, was concerned with the problem of demarcation. Kant sought to find the boundary of what can be called 'objective knowledge' in order to separate it from speculative metaphysics. (Popper drew a demarcation between scientific and non-scientific disciplines.) Kant's famous *Critique of Pure Reason* (^{2}1787) was a sophisticated attempt to confine reason to its proper limits. Reason, if not properly constrained, indulges in flights of fantasy. On the other hand, if reason is chained to mere perceptions, it will never transcend the limits of what is given to it in experience. Kant therefore sought a compromise between Continental Rationalism (Descartes, Leibniz) and British Empiricism (Locke, Hume). Hume's scepticism awakened him from his 'dogmatic slumber'. Kant's problem, according to Popper, was to find the right balance between reason and experience: how mere perceptions can be turned into objective knowledge. Kant's insight, which Popper wholeheartedly accepted, was that our 'perceptions' are already interpretations; 'interpretations of (…) the whole situation in which we find ourselves when 'perceiving'….'. (*C&R* 1963: 387) Kant postulated that human minds are equipped with an a priori apparatus whose job is to convert subjective perceptions into objective judgements. This apparatus consists of categories—causality, conservation of substance, space and time—which are fixed and unchanging. These underlying a priori categories make objective scientific knowledge, such as Newton's mechanics, possible. Popper employed the term 'decoding' to explain how this works:

> I show in my paper [Two Faces of Common Sense 1970, in *Objective Knowledge* 1972: Chap. II, §§3, 13] that if we start from a critical commonsense realism (…) then we shall take man as one of the animals, and human knowledge as essentially almost as fallible as animal knowledge. We shall suppose the animal senses to have evolved from primitive beginnings; and we shall look therefore on our own senses, essentially, as part of a decoding mechanism – a mechanism which decodes, more or less successfully, the encoded information about the world which manages to reach us by sensual means.

It was Popper's 'fallibilism' which drove a wedge between his and Kant's views. He continues:

> There is some reason to think that our senses and our brain operate fairly well together in this business of decoding, but no reason at all to allow them or us any "direct" knowledge of anything immediately "given". ('Replies' 1974: 1059-60)

A little later he adds that sense organs are part of 'our expectational apparatus. This apparatus works well, by and large, astonishingly well, although it is far from perfect: it is *fallible*'. (*Ibidem* p. 1112; italics in original)

Popper accepted that our minds take an active part in the formation of knowledge. But Kant's a priori categories can be interpreted in a psychological sense, which Popper rejected, or a temporal sense, which he endorsed. That is, he agreed with Kant that the human mind imposes laws on nature 'but we rarely succeed'. (*Objective Knowledge* 1972: 68) What did he mean? In his more extreme moments Popper claimed that 99% of our knowledge is a priori in a logical sense. (*Propensities* 1990: 46) He did not agree with rationalists, such as Descartes and Leibniz, that we possess inborn *knowledge* of the external world. But he modified this rationalist thesis. It is still Kant's a priori but in a temporal or genetic sense. (*Propensities* 1990: Chap. II; *Offene Gesellschaft* ^{3}1983: 45, 65–70; *Zukunft* ^{4}1990: 29–31)

> All this led me to the view that conjecture or hypothesis must come before observation or perception: we have inborn expectations; we have latent inborn knowledge, in the form of

latent expectations, to be activated by stimuli to which we react as a rule while engaged in active exploration. All learning is a modification (it may be a refutation) of some prior knowledge and thus, in the last analysis, of some inborn knowledge. (*Autobiography* 1974: §10, p. 40)

At the beginning of *Logik* Popper argued against the positivists that all our experience—even the perception of a glass of water—is theory-impregnated. We cannot acquire experience without some prior form of knowledge. Even a shopping list is a primitive form of theory, which tells a shopper what to look for on the supermarket shelves. In science this procedure is more striking because theories encourage scientists to look for unexpected phenomena. For instance, if theory had not instructed cosmologists to look for dark matter and energy, they could not have hoped to find it through 'mere observation'. Popper then sided with Kant when he asserted that experience and observation are theory-guided. He also agreed with Kant's thesis that theoretical claims without the test of experience remain mere speculations. But he parted company with Kant in a dual sense: (1) He replaced Kant's set of fixed unchanging categories with the more concrete theoretical constructions of the natural sciences. These theoretical constructions remained conjectural. (2) Einstein's replacement of the Newtonian notion of gravity by four-dimensional space–time convinced Popper that science undergoes a constant process of revision and renewal. This was strikingly brought home to the young Popper when Eddington famously confirmed, in 1919, Einstein's prediction of the bending of light near gravitational bodies (see Fig. 2.1a).

With these provisos in mind, Popper accepted Kant's famous dictum that 'thoughts without content are empty, intuitions without concepts are blind'.

Kant's philosophy is traditionally interpreted as a synthesis between rationalism and empiricism. Whilst neither rationalism nor empiricism in themselves suffice to explain the objectivity of knowledge, their synthesis does. When Popper presented himself as a rationalist, he meant it in a Kantian not a Cartesian or Leibnizian sense.

Enlightenment thinking on science was one bond that indebted Popper to Kant. He was also impressed with Kant's political philosophy, in particular his cosmopolitanism. This was Kant's vision of a confederation of nation states for the purpose of securing peace amongst the nations of his time. Kant argued that civil order and the rule of law could be upheld only within a state if it was respected between nations. He envisaged international organizations, like the UN or the EU today, to encourage trade between peoples and thus to reduce the risk of conflict. In his period as a public intellectual, Popper repeated his opposition to all forms of nationalism, which he had expressed in his early essay on the notion of 'Heimat' (1927) and in *Open Society*.

Popper's thought was dominated by his defence of objective knowledge, which was accompanied by an increasing interest in evolutionary biology. It encapsulated an evolutionary theory of the growth of science, a modification of Darwinism and the distinction between Worlds 1, 2 and 3. From this interest it was but a short step to one of the last remaining mysteries, the emergence of consciousness and the mind–body problem. Evolution also evokes the passage of time. The notion of time occupied a central place in Popper's more elaborate notion of realism.

Chapter 8
Popper's Realism and His Theory of Time

Is it not true that reality must have a definite structure in order that we can speak about it?
(K. Popper, *Conjectures and Refutations* 1963: 213)

Popper often stressed his realist credentials. What did he mean by 'realism'? In the development of his thought an evolution of his notion of realism can be observed. At first his realism seemed to amount to no more than the expression of a belief in the independent existence of the external world; a view, which he described as common-sense realism. The term 'realism' hardly figured in his *Logik der Forschung* (1935) except for a suggestion that realism encompasses a belief in the laws of nature. But his *Autobiography* retrospectively called *Logik* a 'realist book'. However, the notion of realism came to be discussed increasingly in his later work, where it took on a more definite meaning. Metaphysical positions, such as realism, were defensible. In his replies to his critics, he repeats his position:

> Perhaps I may repeat here that I am a realist who makes the following concession to idealism: all our knowledge is theory, and our theories are our inventions, as Kant saw; we try to impose them upon the world. Where Kant was wrong, under the influence of Newton's success, is that he did not see how often we err, and fail to impose our theories on the world. ('Replies' 1974: 1183 [fn 31])

Not only are our theories fallible; our sense impressions are fallible too. In his paper 'Two Faces of Common Sense' (*Objective Knowledge* 1972: Chap. 2) he established a connection between his fallibilism and his *critical* common-sense realism. The latter does not assert that we see the world as it is. Human senses, like animal senses, have evolved but are prone to make mistakes. They are part of the afore-mentioned 'decoding mechanism'. This mechanism expresses the theory-dependence of all observations. Apart from the belief in the independence of an external world, Popper proposed at one point that 'kickability' may serve as a criterion for realism: 'we call physically real what is "kickable" (…)'. (*Quantum Theory/Schism* 1982/1992: 46) This criterion refers to our ability to exploit or manipulate physical processes, for instance for technological reasons. The rise of mercury in a thermometer indicates temperature. Windmills and water dams can be used to

© Springer Nature Switzerland AG 2022
F. Weinert, *Karl Popper*, Springer Biographies,
https://doi.org/10.1007/978-3-031-15424-9_8

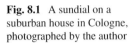

Fig. 8.1 A sundial on a
suburban house in Cologne,
photographed by the author

generate electricity. A sundial exploits the apparent motion of the sun to indicate
time (Fig. 8.1).

It has the unfortunate consequence of making the real dependent on technological
progress. But electrons were real before they could be 'kicked' in experiments.
Besides many things are regarded as real, without being kickable: from the motion
of the planets, the accelerated expansion of the universe, the Andromeda Galaxy,
the Big Bang to, perhaps, the multiverse. More in line with Popper's thinking is the
idea that testability is evidence for realism about the external world. For a theory
which survives repeated severe tests, has some claim to the conviction that it captures
the structure of the physical world. His emphasis on testability explains perhaps his
scepticism towards the Big Bang. (*Zukunft* [4] 1990: 69–70; *Offene Gesellschaft* [3] 1983:
48–50) Theories need to be measured against the physical world; it decides whether
the theory is right or wrong. It is the very independence of the external world, which
allows the severe testing of theories.

A more mature notion of realism emerged in Popper's work after the publication
of *Logik* and it had two essential components: (1) a reference to the structure of
reality and (2) a reference to the reality of time.

8.1   Realism and the Structure of Reality

Let us first concentrate on his notion of 'structural realism'. *Logic* (Appendix *X)
offers a discussion of the nature of natural laws. In contrast to a Humean-type regularity account, he held that laws of nature express *structural* properties of physical
systems. He distinguished natural from logical necessity, accidental from universal

generality. Natural necessity makes no reference to possible worlds but simply means that the physical world is governed by 'structural principles', which exist independently of boundary conditions. If the laws of nature express structural properties of physical systems, such structural properties are not accessible to direct observations. Rather they must be inferred from experimental or observational data. Kepler, for instance, inferred his third law—$A^3 \cong P^2$—from Brahe's observational records. It is an underlying regularity which governs the observable behaviour of planets.[1]

This realization that the appearances do not exhaust the inventory of the physical world became one of the dominant themes of Popper's later philosophy. The distinction between 'appearance' and 'reality' goes, beyond Kant, as far back as the Greeks. In a posthumously published collection of essays—entitled *The World of Parmenides* (1998)—Popper discussed this distinction at length.

> (Parmenides) was the founder of the tradition that all cosmology and all science are a search for the hidden reality, the Thing in itself behind the world of appearances... (*Parmenides* 1998: 126; cf. 200–4)

This Kantian theme ran through Popper's work like a thread, as the following quotes show:

(a) ...our theories make assertions about structural or relational properties of the world; and that the properties described by an explanatory theory must be, in some sense or other, deeper than those to be explained. (*Objective Knowledge* 1972: 197)

(b) I wish to point out that we do not know what matter is, though we know quite a bit about the physical structure of matter. (*Self/Brain* 1977: 175)

(c) We often explain the law-like behaviour of certain individual things in terms of their structure. (*Realism/Aim* 1983/1985: 138)

(d) Like all genuine scientists and seekers after truth, Kepler looks for the reality behind the appearances: hypothetical reality is supposed to explain appearances. And like all seekers after truth he makes many mistakes. But he learns from his mistakes as few others have done. (*Alles Leben* 1996: 146; *All Life* 1999: 74)

(e) My whole methodology consists in the realization that the natural sciences try to find a *reality hidden behind the appearances*. (*Alles Leben* 1996: 150; *All Life* 1999: 77; italics in original)[2]

In a discussion with Magee about Schopenhauer's take on the Kantian 'thing-in-itself', Popper interpreted Schopenhauer as affirming that 'it is knowable (erkennbar)' (16/06/88) [549.19]. Underlying the world of appearances there exists a

---

[1] If a planet, P, takes eight years to orbit the sun, then its average distance from the sun, A, is four times the earth-sun distance.

[2] As far as I know Jeremy Shearmur is the only writer who has previously highlighted the importance of the notion of structure in Popper's work. In his book *The Political Thought of Karl Popper* (1996: 125–31) he argues that the notion of structure can also be applied in the analysis of the social world. Shearmur worked as Popper's assistant from 1971 to 1979, doing similar jobs to Miller's. For instance, he compiled the Index for *Objective Knowledge* and worked with Popper on *The Myth of the Framework*.

'hidden' world, an underlying structure. It is this underlying structure that produces the appearances. This nexus is also at work in Popper's propensity interpretation of probability: the finite, observable frequencies are generated by the underlying propensities (tendencies, dispositions), which constitute a dispositional structure.

Popper did not share the Parmenidean thesis that the world of appearances is a world of illusions. The appearances are as real as the underlying reality. After Newton the world of appearance became the world of scientific truth. (*Parmenides* 1998: 134 [fn76], 144) Popper distinguished surface and depth reality. The underlying structure does not just coexist with the surface reality; rather the observational world is to be explained by 'structural universal laws'. Laws of nature express the hidden structure of the surface reality. All science is a 'search for this hidden reality'. There is thus a link between the appearances and the underlying structures: the latter are attempts to explain the former. The unknown (structural regularities) explains the known (phenomena). The function of scientific theories is to explain the phenomena through the corresponding realities behind them.

> Thus, scientific explanation, whenever it is a discovery, will be *the explanation of the known by the unknown*.[3]

The task of science is made possible because reality has a definite structure. (Appendix *X of *Logic* on natural laws complements the belief in an underlying structural reality). The explanation may be mathematical, as in Kepler's laws or causal, as the explanation of, for instance, rainbows illustrates.

Rainbows are the result of a combination of several processes: (a) refraction of sunlight at the boundary between the air and the airborne water droplets; (b) reflection of light in the interior of the drops; and (c) interference of the refracted and reflected light beam. The angle of deviation, δ, differs for different 'colours' which make up sunlight. Observers see a rainbow on the opposite side of the sun in the sky, the anti-solar point. When the sun is high in the sky, the curvature of the rainbow is small but large when the sun is low. The highest curvature in the shape of a crescent is achieved at dawn and dusk, respectively. One can distinguish a primary and a secondary rainbow. A rainbow can be modelled as a circle of angular radius θ and centred on the anti-solar point (Fig. 8.2a).

A primary rainbow has a radius of 42.5° and a width of 1.5° with respect to the anti-solar point.

It is caused through a combination of a double refraction and a single reflection of the light rays in a water drop. Dispersion then causes the widening of the ray with the sequence of colours from red, orange, yellow, green, blue, indigo and violet (from outside to inside) (Fig. 8.2b).

---

[3] See *Objective Knowledge* 1972: 191, italics in original; cf. *Realism/Aim* 1985: Pt. I, Chap. I. §15, p. 132; C&R 1963: 102f, 174; *Open Universe* 1988: 95. In a similar vein, he wrote to Schrödinger (19/12/52) that 'physics is and remains the explanation of the known (or observable) by the unknown (and unobservable). For this reason one has to make the greatest demands on theory (the unobservable) with respect to its consistency. Expediency cannot be an excuse for laxness in this regard.' [347.18; my translation]

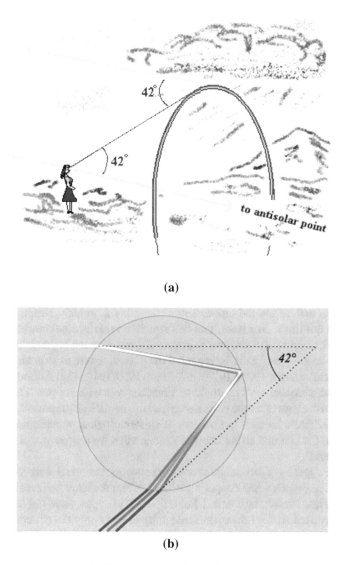

Fig. 8.2 a Anti-solar point and radius of primary rainbow. *Source* http://eo.ucar.edu/rainbows. **b** Double refraction and single reflection in a water drop, giving rise to a primary rainbow; $\delta = 42°$. *Source* Wikimedia Commons

Popper's insistence on *structure* is an important part of his view of science. It is reminiscent of the recent debate about 'structural realism' in the philosophy of science. It started with an influential article by Worrall (1989) but has its roots in earlier positions. I asked Steven French, of the University of Leeds, as a major contributor to this debate, whether Popper's name should be added to the list of forerunners of the modern debate on structural realism. Today's structural realists refer to thinkers such as Ernst Cassirer, Arthur Eddington, Henri Poincaré, Bertrand Russell and Hermann Weyl but they forget Popper. He suspects that 'Karl Popper may be a new figure in the history of structuralism who has been overlooked'. (Interview with Steven French 18.06.2019)

To give a brief indication, structural realists divide into those who claim that 'all there is, is structure', such as French (this position is called 'ontic structural realism'), and Worrall who holds that 'all we know is structure' (epistemic structural realism). In both cases the emphasis is on the structural relations between entities and not on the nature of entities themselves. Depending on which version of structural realism is adopted, they are considered to be either unknowable, derivative or even non-existent. Ontic structural realism is a metaphysical view about the furniture (ontology) of the world. It claims that primarily only structures exist and entities can be derived from these structures. Epistemic structural realism is a claim about the extent of our knowledge about the external world. It holds that our epistemological access to the natural world is limited to structural features. It is agnostic about the question whether there may be some non-structural reality besides the structural reality. In fact, it holds that when there is theory change, there is often continuity at the structural level, but not at the level of non-structural entities. Popper clearly thought that underlying the 'appearances' there lay a 'hidden' structure of reality. But he did not think, like Kant, that this structure remained unknowable. Rather it was the task of science to discover it. Popper did not go much beyond this stipulation and some illustrations.[4] Still, he regarded the laws of nature as structural properties 'of our world'. (*Logik* 1935/[5]1973: 387; *Logic* 1959/1980: 432) Although structural realists have a broader view of structure, French agrees with Popper: 'I try and argue, exactly, that the laws of nature as characterized mathematically represent the structure of the world'. And he adds in criticism of epistemological structural realism: 'It is by virtue of the fact that all we know is structure that we should accept that all there is, is structure'.

Would Popper have subscribed to ontic or epistemic structural realism? One difference between French's and Popper's positions is that the latter believes in the reality of appearances. French, by contrast, holds that talk of appearances (tables and chairs) can be dispensed with: 'I think fundamentally there is only the structure that gives rise to the appearances. To take that word "appearance" seriously (table, chairs, us, the planets) [means they] are derivative entities…'. But Popper's criticism of Heisenberg's indeterminacy relations and the Copenhagen view showed that his notion of structure included the acceptance of the existence of particles. It is difficult to say whether Popper would have agreed with epistemic or ontic structural realism. His emphasis on the reality of underlying structures would have made him sympathetic,

---

[4] The ontic structural realists would not use simple examples, such as the rainbow. They are interested in the relationship of ontological structure to the individuality of quantum particles and space-time points; the role of symmetries and models in physics, and the interpretation of quantum mechanics. Popper, on the other hand, was not interested in such ontological questions. But he shared the interest in interpretative issues in quantum mechanics, especially the role of Heisenberg's 'uncertainty' relations.

I think, to the view that the appearances are derivative. Knowledge of the underlying structure often suffices to deduce the observable reality. This idea is also central to Popper's propensity interpretation of probability. On the other hand, Popper was sceptical of reductionism. He embraced the reality of appearances, propensities and particles. Thus, he may have been less sympathetic to epistemic structural realism. Furthermore, there is a major point of difference between Popper and the ontic structural realists. Popper argued for the reality of time, whilst structural realists like French (and many scientists) sympathize with the notion of the block universe. This is the view that time is unreal.

Neither Popper's insistence on the reality of structure nor his dynamic view on the reality of time has received enough attention in the literature (in my view). Yet they are major parts of his philosophy of science.

8.2   The Reality of Time

The second feature of Popper's more mature view of realism is his belief in the reality of time. In his *Autobiography* (1974: 103; cf. *Open Universe* 1982/1988: 2–3 [fn3]) he affirmed that 'the reality of time and change are the crux of realism'. In fact the 'problem of the *reality of change*' is the 'problem of *the reality of time*'. (*Parmenides* 1998: 166; italics in original) These and other statements throughout his published work show that Popper defended what is known as a 'relational view of time'. It was first proposed by the German mathematician and philosopher Gottfried Wilhelm Leibniz. According to Leibniz's formula, 'time is the order of the succession of events' in the universe. The notion of changing events is therefore central to the relational view of time. It stands in stark contrast to Newton's realist notion of time, according to which time is an independent parameter amongst physical properties. In a nutshell it can be said that according to Leibniz the 'universe *is* a clock', whilst according to Newton the 'universe *has* a clock'. Popper argued repeatedly against this Newtonian view, as well as the block universe, which is often associated with the Special theory of relativity. And although he was much influenced by Kant, he also rejected an idealist view of time. (According to Kant's idealist view time is a mental form.) By contrast, the relational view of time is a constructive part of his philosophy. The passage of time is an inference from the objective succession of events. The universe, therefore, *is* a clock, according to Leibniz's formula. There is evidence for this Leibnizian interpretation in Popper's work. In a long essay on 'Invariants' (1965), published in *The World of Parmenides* (1998: Essay 7) he writes that for a realist 'temporal relations between real events are real' even though 'space and time' are abstractions. (Recall that irreversible propensities also determine the direction of time. *Self/Brain* 1977: 26 [fn4]). Time is a 'legitimate theoretical notion, and so is the direction of time'. ('Replies' 1974: 1141) The very succession of events in the physical universe implies that change is real. But if change is real and time is inferred from the scenario of changing events, then time is a relational concept. Time exists in relation to the occurrence of events and not independently of them. The reality of changing events is central to Popper's realism and his rejection of determinism. A typical statement appears in his work on Parmenides (Essay 7, Addendum) where he affirms that 'all change is due to movement'; the world consists of processes. Leibniz inserted in his formula the term 'order'; he spoke of the *order* of the succession of events. Although Popper did not repeat Leibniz's formula verbatim, he subscribed to it. A changing world, which implies for him the openness of the future, is based on regularities, as it was for Leibniz. The order to which Leibniz referred was therefore a reference to the regular succession of events. For Leibniz it was a Newtonian order; the order of events in classical physics. For Popper, it is an Einsteinian order; the order of events in relativistic physics. This emphasis on regularities, which govern the physical world, refers back to the first aspect of Popper's realism. For Popper emphasized that reality must be explained by universal laws.

Gottfried Wilhelm von Leibniz (1646–1716), philosopher and diplomat. *Source* Wikimedia Commons

But physical laws, it is sometimes objected, do not distinguish between past and future; they make the order of events reversible. It makes no difference to the fundamental laws of physics whether processes run forward or backward in time. The trajectory of a ball looks the same whether it is described, mathematically, from its launch A, at time t_1 or from its landing B, at a later time t_2. A planet also reverses its orbit to return to the initial state it occupied at time t_1. Does this mean that time itself has reversed? No, replies Popper, 'time will have elapsed (…) *without* running back, or returning in the opposite direction, or changing its "arrow"'. ('Replies' 1974: 1141; italics in original) For even though the planet returns to its original position in space, its return will occur later than the beginning of its orbit, as measured by an external clock. If an athlete runs around a circular track repeatedly, s/he will cross the starting line several times. Yet one would not be tempted to conclude that each time the clock reverses. On the contrary, the number of times the runner completes a cycle will be an indication of the *passage* of time. Popper does not only speak of the lapse or *passage* of time. He goes beyond Leibniz by asserting the direction or *arrow* of time. What is the difference?

8.3   The Arrow of Time

As just mentioned, Popper regarded the 'reality of time' as the crux of realism. Between 1956 and 1965 he published four papers on the 'arrow of time' in the prestigious journal *Nature*. In 1967 he continued this discussion on the arrow of time in two further papers, also in *Nature*. He returned to the question of the arrow of time in his book on Parmenides. The theory of time was thus central to Popper's philosophical thought.

When Popper stated that 'time was real' he meant that the reality of time is inferred from the succession of physical events. The succession is not just an ordering of events according to an earlier–later distinction. This succession (before–after) has a direction and this direction constitutes the arrow of time. For change involves irreversible events and processes. Popper agreed with G. J. Whitrow, the author of an influential book on *The Natural Philosophy of Time* (1961):

> I think that the view of those great physicists "who deny that time is 'real' or ... seek to prove that it is a derivative concept of non-temporal origin" makes nonsense of most biological phenomena, including the growth of scientific knowledge. (Whitrow 1961: 311, quoted in *Nature* 1967b: 322)

The reader may ask in which way this endorsement of the reality of time requires a further distinction between the *passage* and the *arrow* time? It is reasonable to make this distinction. We are aware of the passage of events on a local but not on a cosmic scale. In our local neighbourhood we experience a succession of events: order turns into disorder, things decay, the new grows old. Historical evidence tells us that Egyptian pyramids were built before the Eiffel Tower (1889), that the dinosaurs lived before the appearance of the first hominids (around 4 million years ago). There is change all around us. But these local asymmetries of events tell us nothing about the ultimate fate of the solar system and the universe. Humans would still experience the passage of time even if the history of the universe displayed no unidirectional, irreversible pattern. A mouse on a treadmill may think it is going forward in time. It can count the number of times the wheel turns. But to the experimenter the creature is literally going around in circles. In a similar way the universe, as a whole, may reach a maximum point of expansion and then recollapse on itself to a Big Crunch. But local observers in such a closed universe would still see time as marching forward in one direction. (See Weinert 2016: Chap. 14)

When Popper spoke of the reality of time he had both local and cosmic 'arrows' in mind: the passage of time according to the succession of local events, and the irreversible expansion of the universe. But he had a special reason for singling out the cosmic arrow of time. For he objected to a specific interpretation of the arrow of time in terms of the Second law of thermodynamics. The Austrian physicist Ludwig Boltzmann and the British astronomer Arthur Eddington had both identified the arrow of time with the universal increase in disorder. This increase in disorder is a popular, if misleading popularization of the meaning of the Second law. But we can ignore this finer point and focus on Popper's objection to the identification of the arrow of time with an increase in disorder. The Second law asserts that physical systems tend to move from a state of order to a state of disorder, if there is no interference to restore order. This phenomenon is well illustrated in the expansion of a cloud of gas molecules in a sealed container, the scattering of toys in a child's playroom, in the cooling of hot liquids, in the ageing process of plants, animals and humans alike. It is possible to restore order, at least sometimes, by expending energy. One can reheat cold coffee in a microwave oven, parents can restore order in a child's playroom, scientists can return molecules to the corner of the container by moving a piston, and even the lifetime of food can be extended by keeping it in a freezer. But life itself cannot be restored. This universal tendency towards disorder also exists in the universe. Cosmologists observe an expanding universe, even an accelerated expansion. It makes it unlikely that the universe will return to its initial state (Big Crunch). This irreversible expansion suggests that the universe displays a cosmic arrow of time.

Ludwig Boltzmann (1844–1906), Austrian physicist and philosopher. *Source* Wikimedia Commons

At first sight it seemed a good idea to use the universal tendency towards 'disorder' (or the dissipation of energy) as a basis for the cosmic arrow of time. For it is an objective phenomenon, which scientists have elevated to the status of a law. When the Second law was discovered, physicists like Boltzmann and Eddington thought of it as an iron law of nature. Eddington granted the Second law 'a supreme position amongst the laws of Nature'. (Eddington 1932: 74) But both came to realize that the law of increasing entropy is a statistical law—hence it allowed fluctuations. However hard it is to imagine, the laws of physics do not forbid the spontaneous reheating of a cold cup of coffee, in the fullness of time. That is the meaning of a statistical law: it allows exceptions. Even if all our human experience seems to demonstrate that coffee will *never* reheat on its own, even if millions of cups of coffee cool as expected, the laws of physics allow, given sufficient time, that at least one cup may reheat itself. But patience is required. Physicists, for instance, have estimated the amount of time it would take for a volume of gas, containing 10^{18} molecules, to return to its initial state (given by position and momentum variables). It is assumed that each molecule (with an average molecular velocity of 5×10^4 cm/s in both directions) would return to within 10^{-7} cm of each initial position variable and within 10^2 cm/s of each velocity variable. The estimated time for a return to such a configuration would require $10^{10^{19}}$ years, which is well beyond the estimated age of the universe (~ 10^9 years). (See Schlegel 1968: 52–3) According to a more recent estimate, a return of all the particles in a two-chamber system to just one chamber has a probability of $10^{-6x10^{22}}$ and the mean time $\prec T \succ$ for such an occurrence is of the order of $10^{6x10^{22}}$ seconds, the estimated age of the universe. (See D'Abramo 2012)

Popper's intervention in this debate about the cosmic arrow of time had two purposes: (a) he objected to the identification of the arrow of time with entropic processes; (b) he pointed out that there are non-entropic processes, which reveal the irreversibility of temporal phenomena.

These two points are linked. Once it is realized that the Second law is a statistical statement, it becomes difficult to uphold the *identification* of the arrow of time with the law of entropy increase. For one thinks of time as unidirectional, not subject to 'reversals' or fluctuations. As Popper observes (*Nature* **207**):

> ...the 'arrow' of time or the 'flow' of time does not seem to be of a stochastic character: nothing suggests that it is subject to statistical fluctuations, or connected with a law of large numbers. (Popper 1965: 233)

Objecting to Boltzmann's identification of the arrow of time with increase in entropy, he adds (*Nature* **177**):

> The suggestion has been made (first by Boltzmann himself) that the arrow of time is, either by its very nature, or by definition, connected with the increase in entropy; so that entropy cannot decrease in time because a decrease would mean a reversal of its arrow. Much as I admire the boldness of this idea, I think that it is absurd, especially in view of the undeniable fact that thermodynamic fluctuations do exist. (Popper 1958: 403)

Popper therefore considered a statistical theory of the arrow of time to be unacceptable. But the problem is not so much the relevance or irrelevance of statistical-thermodynamic phenomena for time. The problem is the *identification* of the arrow of time with increasing entropy. For entropic fluctuations exist but not temporal fluctuations. Temporal processes are irreversible. The relevance of the Second law would change if increasing entropy could serve as an *indicator* of the arrow of time. If it were regarded as one of many indicators, from which the arrow of time could be inferred. Such a reassessment of entropy is in line with the relational view of time. Entropy is a mere criterion, rather than an identifier of the arrow of time. Consider an analogy: fever is an indicator of many different diseases, but it would be a mistake to identify fever with one particular illness. Mere fever could easily lead to the wrong diagnosis. Popper criticized Boltzmann for the use of the Second law—the law of entropy—in the theory of time. He also disagreed with Schrödinger on the usefulness of entropy as a measure for the reality of time [347.18]. In his *Nature* articles he suggested that other considerations should be taken into account.

There are many criteria from which an arrow of time can be inferred, apart from entropic increase: the formation and development of the solar system, the accelerated expansion of the universe, the evolution of biological species, and the measurement process in quantum mechanics. But Popper concentrated on a much more mundane example. Contrary to the common view that classical mechanics 'can describe physical processes only insofar as they are reversible in time', he produced what he called a 'trivial' counterexample, his pond analogy: a stone is dropped into the centre of a lake. As everyone knows, waves will diverge towards the shore of the lake. From the divergent waves it can be inferred that some local disturbance caused the ripples on the lake's surface. A stone hit the middle of the lake. Could the waves have originated on the shore of the lake? Could this process be reversed? It is highly unlikely that conditions on the lakeshore will conspire to produce converging waves towards the centre, sufficient to lift a stone from the bottom of the lake. Imagine a film is shown of the event: viewers will judge it as typical behaviour that the stone causes divergent waves travelling towards the lakeshore. What if the film were shown in reverse order? 'Contracting circular waves of increasing amplitude' would not be regarded as 'a possible classical process'. Why is this so? The edges of the lake are so uneven that they are extremely unlikely to produce converging waves towards the centre of the lake. This thought experiment of the reversed film can be applied to many phenomena: a broken vase rising up from the floor to re-assemble itself on the table; a rotten piece of wood regaining strength. In all these cases everyone would be instantly convinced that the film was running in reverse. Why? Because everyone knows that the smooth initial conditions of the process under consideration are markedly different from the fragmented final conditions (the broken pieces, the uneven shore). (But a film of a ball running around a circular track in either direction would not allow viewers to distinguish between forward and reverse motion.) Popper therefore concludes (*Nature* **177**):

> Thus irreversible classical processes exist. (On the other hand, in statistical mechanics all processes are, in principle, reversible, even if the reversion is highly improbable.) Although the arrow of time is not implied by the fundamental equations, it nevertheless characterizes most solutions. (Popper 1956a: 538)

Popper's contribution has been praised as 'advancing our understanding of time's arrow by calling attention to the role of non-entropic *de facto* irreversibility'. (Grünbaum 1974: 795; italics in original)[5]

Popper therefore argued that time is unidirectional—it has an arrow. Our notion of time is an inference from the physical changes that are observed on local and cosmic scales. And if it is accepted that there is an arrow of time, it follows that there is a passage of time. In other words: a cosmic arrow implies local arrows. But a cosmic arrow cannot be inferred from local arrows. Independent evidence is required to establish a cosmic arrow of time. For Popper, then, the reality of change implied the reality of time. It also implied indeterminism, a view, which Popper had defended since the 1950s. We should assume that the future is open: 'it is alterable (…) and partly foreseeable'. (*Parmenides* 1988: 175)

---

[5] Grünbaum corresponded with Popper about the arrow of time and visited him at Fallowfield in August 1968. He invited Popper to come to Pittsburgh but Popper declined for personal reasons: he objected to the presence of a certain person there. He did not name this person but from his fallout with Bartley and their correspondence it is clear that Popper did not want to see Bartley who was in Pittsburgh from 1967 to 1973.

The reality of time became a central feature of Popper's realism. The arrow of time is also an objective arrow for it does not depend on human awareness. In Popper's hand the Leibnizian dictum shifted from the succession of events to an emphasis on the reality of change. Change is essentially due to movement. As Popper emphasizes in his article 'Beyond the Search for Invariants' (1965; *Parmenides* 1998: Essay 7) the world consists of processes, not things. As by now he had come to realize that metaphysical theories are criticizable, if not refutable, he offered these reflections on motion and change also as arguments against the ideality of time. That is the view that time is an illusion and that physical reality is timeless.[6] To claim that the passage of time is a human illusion, even if a persistent one, as Einstein once said, is paradoxical: for the illusion of change is still a real event.

Everyone who upholds a Parmenidean view of objective reality (a changeless block universe) must accept a subjective theory of time. It makes time, and change, products of our minds. But the illusion of change is, in its turn, a *real* illusion: we *do* experience change; even if change happens in the illusion. But this means that our consciousness does, in fact, experience change. How can we accommodate this change in an objectively changeless world?

To Popper it seemed to reveal an insoluble problem, or at least a pseudo-problem. He concludes:

> If there is *any* change in the world, even changing illusions, then there is change. (A cinema film exists all at once, but in order to create in us the illusion of motion or change, it must run through a projector – that is, it must move, and change.) And if there is change in the world, then Parmenideanism has to be given up, or radically adjusted. (*Parmenides* 1998: 176; italics in original)

What this argument establishes so far is that change happens at least in the human mind. But Popper wished to claim that there is also change in the physical world. To complete the argument he needed to add that we only experience change *if* our minds observe changes in the physical world. Without this contact with the real world, our minds would be idle wheels. Popper's realism completed the argument: it does assert that the physical world is one of changing processes. The mind decodes these changing events. How the human mind makes contact with the physical world became the focus of his later years. It led to his distinction of the three worlds: the physical world (World 1), the mental world (World 2) and the world of objective thought products (World 3). The interaction between these three worlds resulted in a view called *interactionism*, which had matured since the 1950s. It was his contribution to the mind-body problem. His theory of time goes hand in hand with his emphasis on the evolution of consciousness and the evolutionary growth of knowledge. It was almost inevitable that he would turn to Darwinian biology.

---

[6] This view has a long tradition, but it became popular with physicists in the wake of Einstein's theory of relativity; see my books *The Scientists as Philosopher* (2004) and *The March of Time* (2013).

Chapter 9
Biology, Evolution and Life

Organisms are problem solvers and explorers of their world. (K. Popper/J. Eccles, *The Self and Its Brain* 1977: 138; cf. [321.04])

Popper's engagement with biology harked back to *Logik* (1935) and his participation in a meeting of the progressive Theoretical Biology Club, of which Wiesner and Woodger were both members. His debate with the Post-rationalists drove him towards evolutionary thinking, into the arms of Darwinism. It started with his evolutionary view of the growth of scientific knowledge. But then Popper felt he could make a 'modest' contribution to Darwinism in order to address 'shortcomings' in Darwin's theory. His aim was to explain the apparent goal-directedness of evolution. It was meant as an improvement of the Darwinian theory. This interest in evolutionary theory then led Popper, in cooperation with John Eccles, to apply evolutionary thinking to the age-old 'mind-body' problem.

9.1 On the Scientific Status of Evolutionary Theory

At first, however, Popper took a rather dim view of the scientific status of evolutionary theory. Following on from *Grundprobleme*, his *Logic of Scientific Discovery* (§30) contained a brief footnote on the natural selection of theories by the method of trial and error, without mentioning Darwin. But such an evolutionary approach to the evolution of ideas is different from the question of the scientific status of the Darwinian theory. Popper, at first, held that 'Darwinism is not a testable scientific theory but a *metaphysical research programme*—a possible framework for testable theories'. (*Autobiography* 1974: 134; italics in original) It is applied situational logic, that is, trial and error applied to life. Much of Darwinism was a 'logical truism'. (*Objective Knowledge* 1972: Chap. II, §16; Chap. VI, §18) He meant that, by definition, only those individuals survived that are well adapted; they are the 'fittest'. But fitness is measured in terms of survival rates. So the surviving organisms are the 'fittest', and the 'fittest' are the survivors. But later he retracted the view that

© Springer Nature Switzerland AG 2022
F. Weinert, *Karl Popper*, Springer Biographies,
https://doi.org/10.1007/978-3-031-15424-9_9

Darwinism was a mere tautology and declared that it can become a testable theory. He even conjectured that Darwinism was right. (*Myth* 1994: Chap. 1, §V, p. 9; *Offene Gesellschaft* [3] 1983: 53–5) Popper became, in his own words, 'an admiring but reluctant Darwinian'. (*Knowledge* 1994: 34) He even labelled Darwin a 'fallibilist' and 'falsificationist' [345.13]. This change of mind was reflected, at first tentatively in his *Autobiography* (1974: §37), then more affirmatively in a lecture he delivered at Darwin College in Cambridge in 1977.[1] In his Cambridge lecture he admitted that there were some tests of natural selection, such as industrial melanism[2], although 'severe tests' were difficult to find. He pointed out that the principle of natural selection alone cannot explain all aspects of organic life. For instance, the antlers of deer and the peacock's tail have no direct survival value. On the contrary such elaborate ornaments would hinder the animals in their 'struggle for existence'. Darwin accounted for these phenomena by stipulating an additional principle: sexual selection. These features help in the fight for reproduction. It is nowadays regarded as a form of natural selection. There always loomed, according to Popper, the danger of tautology. But if the theory of natural selection can be formulated in a non-tautologous manner, it becomes testable. In a letter to the *New Scientist* (1980) he was adamant that the 'historical sciences have in my opinion scientific character: their hypotheses can in many cases be tested'. One way of avoiding the tautology is to characterize 'fitness' not in terms of survival rates but in terms of 'surviving-favouring attributes'. As Ernst Mayr (2001: 118) put it: 'To be fit means to possess certain properties that increase the probability of survival'. Stephen J. Gould (2002: 369[fn]) provided an illustration: 'The speediest deer can be specified beforehand, and their differential survival in a world of wolves can then be tested empirically'. (Cf. Lewontin 1972)

In his *Autobiography* Popper reflected on the explanatory and predictive abilities of Darwinism. He did not have Darwin's original formulation in mind, which he still regarded as 'almost tautological', but the neo-Darwinian Modern Synthesis, which combines Darwin's mechanism of natural selection with molecular biology. But even this theory he deemed 'metaphysical' since it is not *strictly* testable. (*Autobiography* 1974: 136-7) It did not strictly 'explain' the evolution of different life forms on earth, although it implied it. Popper drew a distinction between 'explanation in principle' (as it occurs in evolutionary theory) and 'explanation in detail' (as it happens in physics). (*Autobiography* 1974: 138, 179, [fn285]) Evolutionary theory predicts that evolution will be 'gradual'—imperceptibly gradual in Darwin's phrase—but this remains its 'only prediction'. It lacks universal laws, yet it is revolutionary because it explains apparent design in physical terms. (*Objective Knowledge* 1972: 267) As the only serious contender, it was 'invaluable'.

---

[1] 'Natural Selection and the Emergence of Mind' (1977). Magee reassured him that it was a success (14/11/77). He listened to the BBC radio version of the lecture and found that it was 'tightly knit and universally interesting' (06/03/78) [323.01]. Not so Popper. In his reply (09/03/78) he confessed that 'I was so badly repelled by my BBC Darwin lecture that I went out of the room where Hennie listened; and she did not like it any better' [323.02].

[2] Industrial melanism refers to colour changes in various species, for instance moths, as a result of environmental pollution.

Popper underestimated the power of evolutionary theory. Standard textbooks discuss the testability of natural selection and the predictability of evolutionary events.[3] Whilst Popper usually condemned scientism, physics envy and the 'aping of the natural sciences by the social sciences'[4] in these passages of his *Autobiography* he seemed to compare biology unfavourably to physics. This did not stop him from trying to offer what he considered 'some slight improvement of Darwinism (of the theory of the Baldwin effect)'. ('Replies' 1974: 1191 [fn150])[5] At first, he dubbed this 'slight improvement' the *Spearhead model of evolution* [272.05]. But in the 1980s he proposed a distinction between active and passive Darwinism. As Jon Hodge and Greg Radick, of the University of Leeds, told me the term 'spearhead model' is not well known amongst evolutionary biologists. (Interview 16/07/2019) By contrast, the Baldwin effect is a well-known phenomenon today. It is named after the American psychologist James Mark Baldwin who published the idea in 1896. It was Julian Huxley who in his book *Evolution: the Modern Synthesis* (1942) recognized the importance of this 'neglected effect' and promoted it as part of the Modern Synthesis. The Baldwin effect, which Huxley called 'the principle of organic selection', describes the effect of learned behaviour on evolution. Baldwin did not invoke the Lamarckian principle of acquired characteristics but suggested that adaptations can arise from plasticity, that is adaptable behaviour.

9.2 Popper's Contribution to Evolutionary Theory

Popper's shift to evolutionary biology was the result of his long-term interest in the (evolutionary) growth of scientific knowledge. His theory of evolutionary growth is

---

[3] For instance, Darwin provided evidence for natural selection in terms of anatomical comparisons: many species have similar structures (like the forelimbs of vertebrae), which is due to their common ancestry (homology). Other organs have similar functions (like bat and bird wings) but differ in structure or development (analogy). These features are caused by similar responses to environmental pressures. The eye evolved independently in at least forty different species. Further evidence is provided by industrial melanism, and genetic changes in the lineage of species. The adaptation of the male guppy to challenges in his environment—in the presence of predators the male guppy changes its colour pattern—was established by a striking experiment. The gradualism of evolution is not the only prediction of evolutionary theory. Further predictions cover the speciation or the splitting of lineages under specific geographic conditions and the existence of transitional forms. Jerry A. Coyne provides many more examples in his book *Why Evolution is True* (2009).

[4] Cf. *Myth* (1994):75; *Poverty* (1960): 60, 105–6; *Objective Knowledge* (1972): 186; *Zukunft* (1990): 47–49; *Auf der Suche* (1987): 53.

[5] In his writings Popper referred both to the Baldwin effect and 'orthogenesis'. (*Objective Knowledge* 1972: Chaps. 6, 7) The reference to orthogenesis may lead to confusion. It can be understood either in the discredited Lamarckian sense of an evolutionary trend towards perfection or as the apparent goal-directedness of evolution. The French naturalist Jean-Baptiste Lamarck thought of evolution as a ladder of ascent from lower to higher forms of life. Popper's main focus was on 'orthogenetic trends' in the sense of apparent goal-directedness, which can be simulated by Darwinism. (*Autobiography* 1974: 138, 141) He proposed his Spearhead model of evolution, which is based on the Baldwin effect, to account for such trends.

based on the tetradic scheme (here simplified):

$$P_1 \rightarrow TS \rightarrow EE \rightarrow P_2$$

(Problem$_1$ $\rightarrow$ Tentative Solution $\rightarrow$ Error Elimination $\rightarrow$ Problem$_2$).

James Mark Baldwin (1861–1934), American philosopher and psychologist, pictured in 1917. *Source* Wikimedia Commons

He emphasized that organisms, too, are constantly engaged in problem-solving (P_1), trying out different tentative solutions (*TS*) by error elimination (*EE*), engendering new problems (P_2). These problems go beyond the classic struggle for survival and include testing out new behaviour: for instance, a change in food preferences or the conquest of new environments. Julian Huxley also appealed to the conquest of new habitats, i.e. the Baldwin effect.

> This extension of habitat may in the first instance be dependent on a non-inherited modification of behaviour, mutation and selection later stepping in to fix the change genetically (the "organic selection of Baldwin and Lloyd Morgan..."); or genetic variants may find themselves in surroundings to which their constitution is better adapted than was the normal environment of the species (pre-adaptation...). In either case, migration will have been advantageous to the species as well as to the individual. (Huxley 1942: 114)

In his book *The Walking Whales* (2014) the Dutch-American palaeontologist Hans Thewissen described such an extension of habitat: the migration of whales from land to water in a span of eight million years. His findings are based on concrete first-hand fossil evidence gathered in India, Pakistan and other parts of the world.

Popper believed that he could improve on Darwinism, especially the problem of orthogenesis. What was wrong with Darwinism in his eyes, apart from its near-tautological status? In *Objective Knowledge* (1972: 269f) he claimed that the 'real problem for Darwinism is to explain what looks like goal-directed evolution'. That is, how a complicated organ like the eye can result from accidental, independent mutations. (1972: 272ff) He called this the problem of 'orthogenesis', which he characterized as 'sequences of evolutionary changes in the same "direction"'. (*Auto-biography* 1974: §37) The evolution towards complexity and variety was not a new problem. The English evolutionist Thomas Huxley—known as Darwin's bulldog—already proposed a solution in terms of natural selection. Evolution, according to Darwin and Huxley, is a slow process, imperceptibly slow. It occurs as a result of genetic mutations and the challenges the environment posed to the individual organisms. Organisms not only face the threat from predators but also competition for food from members of the same species. Darwin accepted the Malthusian view that offspring always outrun available food resources. Organisms with a slight advantage (better eyesight, faster legs and sharper teeth) over others will tend to survive better and reproduce more successfully in their particular habitat. Hence they have a better chance of passing on their genes. This process is called natural selection. Evolution results from genetic mutations and natural selection. Evolution is a cumulative process in that it preserves the advantages it has bestowed on the members of a species. Popper was sceptical that natural selection offers a solution to every evolutionary problem. Even if he accepted that it can provide explanations in principle, natural selection cannot account for all evolutionary phenomena. As he explained in his Cambridge lecture (1977, §2):

> The theory of natural selection may be so formulated that it is far from tautological. In this case it is not only testable, but it turns out to be not strictly universally true. There seem to be exceptions, as with so many biological theories; and considering the random character of the variations on which natural selection operates, the occurrence of exceptions is not surprising. Thus not all phenomena of evolution are explained by natural selection alone. Yet in every particular case it is a challenging research program to show how far natural selection can possibly be held responsible for the evolution of a particular organ or behavioural program.

In a way these reflections on Darwinism harked back to *The Poverty of Historicism* (1957/²1960: §27). Popper denied that there can be a (universal) *law* of evolution. The evolutionary hypothesis was a particular 'historical statement about the ancestry of a number of terrestrial plants and animals (…)'. It was the description of a unique historical process. But evolutionary theory does not account for the origin of life. Darwin agreed. Nor does it explain, Popper continued, orthogenetic trends, trends towards higher forms of complexity. It is true that Darwinism dismisses the Lamarckian view of evolution as an inevitable trajectory towards higher forms of life. It explains the complexity of organs (like the eye) as an adaptation to environmental niches. Popper rejected this image of organisms as *passive* actors in a changing and challenging environment. He depicted organisms as *active* problem solvers; hence his appeal to the Baldwin effect, which he mentioned several times in *Objective Knowledge* (1972: 149, 245 [fn56], 268 [fn11], 269, 270, 272). In *The Self and Its Brain* (1977: 12) he labelled the effect 'organic evolution'.

> The theory of organic evolution starts from the fact that all organisms, but especially the higher organisms, have a more or less varied repertoire of behaviour at their disposal. By adopting a new form of behaviour the individual organism may change its environment.

His example was a change in food preferences, which allowed the organism to exploit new ecological niches.

By this individual action, the organism may "choose", as it were, its environment; and it may thereby expose itself and its descendants to a new set of selection pressures, characteristic of the new environment.

He criticized Darwinism for having underestimated cases, in which habits change first. The emphasis shifted to the Baldwin effect. In his Tarner Lecture (at Trinity College, Cambridge, October 1956) Schrödinger (1958), following Huxley, also distinguished passive from active Darwinism, without explicitly naming the Baldwin effect.

In our conversation, Jon Hodge summarized the Baldwin effect as follows:

What is in common between Baldwin and Popper (...) is that the change in behaviour of the animals causes genes to be advantageous that were not previously advantageous. They spread through the population. It causes some genes that were neutral or advantageous to be disadvantageous and they are excluded. So you get genetic change, not for a Lamarckian reason but because the selection pressures on the genes are changed by the behavioural innovation.

A species that is threatened by a new predator provides an example. Some individuals of that species may have the potential to display a new type of behaviour which makes it more difficult for the predator to catch them. The individuals who have adopted the new behaviour will have a better chance of survival and reproduction. It causes genetic change in the long run through selective pressure. An organism's ability to learn new behaviours will affect its reproductive success and will therefore have an effect on the genetic make-up of its species through natural selection. It is not Lamarckism because new behaviour is not inherited. Popper's favourite example of this process was the beak of the woodpecker:

A reasonable assumption seems to be that this specialization started with *change in taste* (preferences) for new foods which led to genetic behavioural changes, and then to new skills (…); and that the anatomical change came last. A bird undergoing anatomical changes in its beak and tongue without undergoing changes in its taste and skill can be expected to be eliminated quickly by natural selection, *but not the other way round*. (Similarly, and not less obviously: a bird with a new skill but without the new preferences which the new skill can serve would have no advantages.) ('Replies' 1974: 140; italics in original)[6]

The Baldwin effect appealed to Popper since he always put strong emphasis on the activities of individuals (either as members of a species or as creative problem solvers). But Popper went beyond the Baldwin effect when he claimed that 'the only creative element in evolution is the activity of the organism' ('A New Interpretation of Darwinism', in Niemann 2014: 119). His Medawar Memorial Lecture (1986) shifted the emphasis from passive to active Darwinism.

Popper's 'enhancement' of Darwinism consisted in the following theses:

1. All organisms are involved in troubleshooting through tentative behaviour and error elimination.

---

[6] The reader may be reminded that orthodox Darwinism explains the different beaks of finches in the Galapagos Islands as a result of random genetic changes, which favour some individuals in the struggle for survival over others.

2. The spearheads in the adaptation of the individual are the various behaviour patterns tried out by the individuals. They are like evolutionary experiments, for instance in the conquest of new environments. The species 'employ' the individuals as spearheads. (The emphasis on individuals is not surprising, given Popper's long-standing commitment to 'individualism'.)

3. Biologists have missed that behaviour is more important than anatomy. Everything follows from behaviour: 'habits change first.' (*Self/Brain* 1977: 13) The woodpecker develops new food preferences, which lead to genetic 'behavioural' changes, and then to new skills and new anatomy.

4. Under certain conditions, a behavioural trait may become entrenched through mutations, leading to genetic change in species. (It can also lead to unfitness). The main point is that 'mutations can succeed only if they fall in with already existing behaviour patterns.' (*Knowledge* 1994: 69) What leads in evolution are behavioural changes; what leads to changed behaviour are new aims, skills and only then do anatomic changes occur.

5. Popper rejected the accusation that he had retreated to Lamarckism: behaviour traits are not inherited; they only stimulate a selection pattern. New aims (skills, dispositions) may turn previously unfavourable into favourable mutations, through natural selection.

6. The selection of a mutation will be strongly dependent on adapted behaviour.

In the early 1930 Popper had challenged the Vienna Circle and its dominant philosophy of logical positivism. Fifty years later he used his Medawar Memorial Lecture to challenge the dominant view of neo-Darwinism with its emphasis on random mutations and natural selection. In the presence of influential biologists and Nobel Prize winners, including Peter Medawar and Max Perutz, he argued for a replacement of passive by active Darwinism, i.e. a modification of the Modern Synthesis.

Popper tended to think that his contributions to various areas of philosophy, for instance the philosophy of quantum mechanics, had not been properly acknowledged and appreciated. As we have seen, his propensity interpretation did produce a certain echo amongst physicists and philosophers. How did the Spearhead model or active Darwinism fare? According to Jon Hodge, Popper had rather little influence amongst evolutionary biologists.

> I would guess that if you collected all the papers that have been written about this [Spearhead model] in all the main languages of the world he might have a dozen papers. It has not been a big topic. (…) It is very strange because Popper has many admirers amongst biologists but always for his general theory of falsification'. (…)

And further:

> There are two textbooks on evolutionary biology, which are very valuable as historical sources. They set out the orthodoxy very clearly. And the Baldwin effect gets into both of them but only briefly. Popper will not be mentioned. One of them is by Mark Ridley, the other is by Douglas Futuyma.[7]

Yet views on Popper's contributions to biology are divided. The natural scientist Hans-Joachim Niemann ranks Popper as an eminent philosopher of biology because he proposed a third way between creationism and orthodox Darwinism. This third way is active Darwinism. Niemann traced Popper's interest in biology to the meetings of the Theoretical Biology Club in the 1930s. Following Popper's lead Niemann criticized the 'central dogma' of molecular biology, i.e. that hereditary information can only move from genes to proteins and somatic cells. Niemann deserves credit for having obtained permission to publish Popper's previously unpublished Medawar Lecture, and some other writings on evolutionary biology, in his book *Karl Popper and the Two New Secrets of Life* (2014).

Biologists may be sympathetic, yet critical of Popper's Spearhead model. Max Perutz was an Anglo-Austrian molecular biologist, who shared the 1962 Nobel Prize for chemistry with John Kendrew, for their studies of the structures of haemoglobin and myoglobin. Perutz exchanged some letters with Popper and sent him his 'haemoglobin papers'. They shared the 'old-fashioned view' that 'science is the triumph of reason'. Popper asked him to 'drop the Sir' (26.03.83) [337.13], as he also told Anthony O'Hear (22/03/94) [554.14]. After Popper's lecture, Perutz published an article in the *New Scientist* (1986), in which he offered his assessment of Popper's views on evolution. Perutz wondered whether evolutionary theory needs Popper's revision. He starts out by observing that Popper's rejection of determinism influenced his views on the evolution of species. Popper, he continues, separates Darwinism into an 'active' and a 'passive' part. But he rejects the passive part, which corresponds to the classic Darwinian formula: mutation + natural selection leads to 'higher' forms of life. It is, according to Perutz, a version of philosophical historicism. By contrast Popper emphasized the activity of the individual organisms. They explore their respective environments.

---

[7] M. Riley, *Evolution* (^{3}2004); D. Futuyma, *Evolutionary Biology* (^{3}1998). It is indeed the case that these two books do not mention Popper's Spearhead model, nor do they refer to the Baldwin effect. A random search of other textbooks on evolutionary biology yielded the same result. But Ernst Mayr, who was considered to be one of the world's most eminent evolutionary biologist, mentions the Baldwin effect, approvingly, in his book *What Evolution Is* (2001: 137, 142). He describes it as the idea that 'behaviour is the pacemaker of evolution.' However, he makes no reference to Popper's ideas, although Mayr and Popper corresponded with each other. Mayr appreciated Popper's Cambridge lecture on 'Natural Selection and the Emergence of Mind' (letter of 31/07/79)[325.01]). They were friends, met each other and Popper invited Mayr to Penn.

Max Perutz (1914–2002). *Source* Wikimedia Commons

Perutz then proceeds to examine Popper's two aspects of evolution by using the example of haemoglobin in various animals (camels, llamas, bar-headed geese, deer mice). The haemoglobin molecule in llamas has a higher 'affinity' to oxygen than in camels and enables these animals to inhabit oxygen-poor areas. But this mutation probably took place *before* llamas discovered that they could forage for food in higher regions of the Andes. A similar mutation allows bar-headed geese to fly over the Himalayas in the summer. Haemoglobin-related diseases in humans are a case of mutation of the haemoglobin genes. To be affected, a child must inherit the genes from both parents. Children with the sickle-cell gene are protected against malaria. It is a case of adaptation through natural selection, the passive aspect of evolution. Deer mice, by contrast, may have either of two haemoglobins, which enables them to adapt to different environments. In the case of mice, he finds that 'active' and 'passive' Darwinism will work 'side by side'. 'Darwinian evolution may be either active or passive or a mixture of both'. Perutz agrees with Popper that individuals actively search for better living conditions. But this is only one aspect of evolution. He disagrees with Popper on the role of goal-directed behaviour.

Interestingly, Perutz adds a remark about scientific methodology. Progress, he says, either happens through the hypothetic-deductive or the inductive method. Progress can also come from accidental discoveries without the guidance of a hypothesis. (This latter aspect was also emphasized by Paul Nurse in his *LSE* lecture, 2016).

The jury is still out on whether Popper's active Darwinism really does constitute a third way. There are certainly voices who share Popper's opposition to a narrow interpretation of the Modern Synthesis. It is not a question of the rejection of the Modern Synthesis, but a call for its extension. (Noble 2014; Jablonka 2017)

As far as I can see there are two main reasons why Popper's excursions into evolutionary biology were important. It provided a response to Kuhn's challenge. It inspired him to develop an evolutionary picture of the growth of scientific knowledge with its emphasis on competition and elimination. The other reason is that Popper believed evolutionary theory would provide him with a handle on the 'mind-body' problem, on which he began to focus in the 1970s.

9.3   The Mind-Body Problem

After his extended excursion into evolutionary theory it was almost inevitable that Popper would turn his attention to the problem of how the mind is related to the brain. His *Grundprobleme* (1979: Anhang IX; 2009: Fragments 1933, IX) contained a sketchy discussion of the problem of free will whose 'solution' was inspired by Kant. Popper was at times optimistic that the mind-body problem could be resolved, with the help of evolutionary theory. (*Objective Knowledge* 1972: Chap. VI) At other times he felt pessimistic, declaring that the problem 'will never be solved'. (*Self/Brain* 1977: Preface, p. VII; *Alles Leben* 1996: 72) He nevertheless felt that some kind of 'interactionism', which he had already proposed in the 1950s (*C&R* 1963: Chap. 12), provided at least a conceivable solution. It was a research programme, but complete understanding of the problem would elude us. (*Self/Brain* 1977: 37) In his attempt to grapple with the problem, the French philosopher René Descartes famously proposed an interaction between two substances: the material body, characterized by spatial extension; and the immaterial mind, the seat of thought. The interaction between them was achieved via the 'pineal gland', a small endocrine gland in the brain of most vertebrates. As readers may surmise, this only pushed back the problem. The pineal gland is itself a physical organ and thus the problem of how the brain can affect the mind, via this gland, recurs. Popper and his co-author John Eccles moved away from this substance dualism. Popper noted repeatedly that consciousness is not a substance but a process, an interaction of events.[8] (*Zukunft* [4]1990: 86–7) The supposed interaction between mind and brain took on a non-Cartesian form. It was psycho-physical interactionism with an evolutionary twist.

This view culminated in the publication of *The Self and Its Brain* (1977), a defence of interactionism and emergence. The book was long in the making. The authors started a 30-year-long correspondence in 1945. Already in the early 1950s they agreed on the approach to the mind-body problem—interactionism, which they defended in the book. They discussed the possibility of a book in 1972 and in September 1974 they stayed at the Villa Serbelloni, Bellagio, on Lake Como to plan further details and make recordings of their discussions.[9] In 1975 Plenum Press agreed to publish the book but Popper was still working on his chapters. But a year later Springer published the book. It attracted a lot of attention and was reviewed in both the scientific press (*Nature*) as well as the popular press (BBC Broadcast).

---

[8] Popper did not draw a clear distinction between consciousness, mind and self. He declined to specify what the mind is. (*Objective Knowledge* 1972: Chap. VI)

[9] Peter Medawar and Jacques Monod also visited Serbelloni at around that time. Magee applied to the *Rockerfeller Foundation* to stay at the Villa a year later to work on his Schopenhauer book. Popper's reference (29/01/75) praised Magee as 'an academic philosopher of distinction' and a 'Schopenhauer specialist', who would put Schopenhauer, 'one of the great philosophers' on the map again [323.01]. Note that Popper made several references to Schopenhauer in his writings, starting with *Logic* and *Open Society*.

9.3.1 Interactionism

The mind-body problem is familiar to every self-conscious agent, whether they are trained in philosophy or not. When we feel sad or happy our feelings have an effect on our visible behaviour. Sadness may produce tears; happiness may result in smiles. But how is it possible that mental states, which are immaterial, produce such physical manifestations? The question can be looked at in the opposite direction. How can a purely physical event, like an accident or a promotion, produce such immaterial reactions as sadness or joy? It is one of the outstanding mysteries, which has attracted many provisional solutions: epiphenomenalism, functionalism, materialism or the identity theory, of which Feigl was a proponent. Popper found all these approaches wanting. He criticized materialism for failing to account for World 3 and the higher functions of language, namely the ability to argue, criticize and describe. The mind-body problem and the existence of mental states could not be reduced to mere brain activities. His argument, as he summarized in a letter to Feigl (18/07/66), was that

> …states of a discussion, and similar states of 'content', are no physical states, and need mental states both to produce them and to grasp them (and to make them act upon the physical world) [294.06].

A couple of years later (17/04/73), he shared with Feigl an argument he had heard from Alan Musgrave:

> If my proposal is accepted that two statements describe the same event if and only if they are interdeducible, then a psychological and a physical statement cannot describe the same event [536.10].

Feigl eventually began to have doubts about the identity theory. Popper aimed to solve what he dubbed Compton's problem: how mental products can have physical effects. (*Objective Knowledge* 1972: Chap. VI, §§ 6, 7, 17)

Given such an approach the notion of *emergence* took centre stage. Popper emphasized the 'emergence of the mind' in his lecture at Darwin College, Cambridge (November 8, 1977). Earlier that year he had told Feigl (15/05/77) that he was 'an interactionist' and a believer in emergence as a fact, 'from the point of view of present cosmology' [536.10]. For if mental activity cannot be reduced to neural activity (as both materialism and the identity theory claim) and if the mind is more than a mere by-product of the brain (epiphenomenalism denies that the mind can have an effect on the physical), then both mental and brain states are real. The Cartesian question returns of how they are related without making the 'essentialist' assumption that they are substances.

Emergence basically means that a higher-level state differs qualitatively and quantitatively from the lower-level base state. The emergent state can no longer be reduced to the base state. It is autonomous and follows its own 'laws'. A simple example is the baking of a cake: once the ingredients have been mixed and the dough has gone into the oven, the emerging cake will be qualitatively different from the dough. The finished cake can no longer be broken down into its ingredients.

John Eccles (1903–1997), neurophysiologist and Popper's collaborator. *Source* Wikimedia Commons

By way of analogy Popper held that higher-level mental states were qualitatively different from lower-level brain states. There are two aspects to this problem (*Autobiography* 1974: §39, p. 152): (A) The question of how mental states are related to physiological states or how the mental emerges from the physical. For Popper the mental involves both World 2 and World 3, both subjective consciousness and its objective, autonomous products. But Popper added that this emergence is an evolutionary process, an aspect which was radically absent from the Cartesian account. (B) The emergence of the self and consciousness from evolution. In his Cambridge lecture Popper stated that 'mental powers' were the product of natural selection. 'Mind' he asserted, 'could not have evolved if it had been useless'. Mind and consciousness, therefore, had survival value.

> My central thesis here is that the theory of natural selection provides a strong argument for the doctrine of *mutual interaction* between mind and body or, perhaps better, between mental states and physical states. (…)
>
> I even defend *pluralism*, since I hold that there are three (or perhaps more) interacting levels or regions or worlds: the world 1 of *physical* things, or events, or, states, or processes, including animal bodies and brains; the world 2 of *mental* states; and the world 3 that consists of the *products of the human mind*, especially of works of art and of scientific theories. (Popper 1977: §3; italics in original)

Human consciousness emerges through the interaction of World 2 with World 1 and World 3. 'The mind-body problem must include all three worlds'. (*Knowledge* 1994: Chap. 1) It is between these three worlds that the interaction takes place. Language plays a vital part in this process because it makes objective thought possible. The Self and World 3 evolve with the higher function of language, i.e. the argumentative function.[10] Hence mind and consciousness emerge in stages. The use of language happens of course in World 2, the world of consciousness and mental life. But the invention of abstract theories—their formulation in objective mathematical or theoretical terms and their availability in public records—make them independent of World 2. The products of World 2 activity then migrate to the autonomous World 3, the world of objective mental products. The contents of World 3 have survival value. (*Knowledge* 1994: Chap. 1) Abstract ideas shape human history and help us navigate our course through the natural and the social world. The question nevertheless arises how the interaction between the three worlds takes place and how this interaction relates to Darwinism.

Let us first look at the evolutionary aspect. Popper held that 'body and mind evolve under the pressure of natural selection'. (*Self/Brain* 1977: P3, §20–21) Brain activity was a necessary condition of mental processes. That is, without the evolution of the brain no mental life could blossom. It is a known fact that the brain capacity of hominids increased over millions of years from 400 cm^3 to today's value of around 1350 cm^3. Such a large brain can fulfil functions, which would have challenged the small brains of our hominoid ancestors. From the evolutionary point of view then, 'the self-conscious mind is an emergent product of the brain, just as World 3 is an emergent product of the mind'. (*Self/Brain* 1977: 554; cf. P4, §36; P5) It is a consequence of evolution that consciousness emerged by degrees, as brain capacity increased. (*Self/Brain* 1977: 438; Dialogue II) Evolutionary theory, in fact, leads to very similar solutions.

In his book on human evolution, *The Descent of Man* (1871), Darwin argued that there is no fundamental difference in kind between the mental functions of humans and higher organisms. Complex animals feel emotions and manifest signs of intelligence. According to Darwin, the principle of natural selection is able to explain the evolution of mental and moral faculties. In the idiom of the day, the brain is the organ of mind. But the human brain is a sophisticated and complex organ, which makes it implausible that each of its functions would have been especially selected. For Darwin, natural selection was not the exclusive means of modification. He introduced the principle of sexual selection to explain the origin of features, such as antlers in deer, which seem injurious to the survival prospects of an individual. (Darwin 1859: Chap. IV) Darwin argued that the mental and moral faculties have, most probably, been 'perfected through natural selection', adding, 'either directly or more commonly indirectly'. (Darwin 1871: Part I, Chap. IV, p. 151) A Darwinian can account for non-adaptive modifications by advancing two considerations.

(1) 'Organisms are integrated systems and adaptive change in one part can lead to non-adaptive modifications of other features'. (Gould 1987: 45) Darwin called this the *Principle of the Correlation of Growth*. (Darwin 1859: Chap. V, pp. 143–50; 1871: Part 1, Chap. IV) Slight differences in human body structure lead to

---

[10] Popper added this function to the three functions which one of his teachers at the Pedagogic Institute, Bühler, had identified. Popper made constant references to these functions of language but he regarded his additional argumentative function as the most important.

parallel changes elsewhere in the system. This could explain the doubling of the cranial capacity of the human brain.

(2)   'An organ built under the influence of selection for a specific role may be able, as a consequence of its structure, to perform many other, unselected functions as well'. (Gould 1987: 50) This could explain the evolution of mental capacities. Our large brain may have originated 'for' some set of necessary skills in gathering food, socializing, etc., but these skills do not exhaust the limits of what such a complex organ can achieve. (Darwin 1859: Chap. V; 1871: Part I, Chap. V) 'Natural selection makes the human brain big, but most of our mental properties and potentials may be spandrels—that is, non-adaptive side consequences of building a device with such structural complexity'. (Gould 2001: 104)

Hence for Darwin humans were as much a product of evolution as any other organism. The Darwinians vigorously rejected an appeal to 'new creative forces' to explain human brains, and hence followed the principle of unification that the same phenomena must be explained by the same principles.

But why should Darwinism favour interactionism? What is the biological function of the self? The mind, Popper claimed, helps the adaptation of the organism. (*Self/Brain* 1977: P3, §20) The self-conscious mind exercises plastic control over neural events in the brain. Eccles specified that the controlling role of the self-conscious mind is limited to 'certain neural events', which occur in a specified zone. (*Self/Brain* 1977: E7, §48) There was therefore an interaction between the controlling system and the controlled system. (Popper, *Knowledge* 1994: Chap. 3) This controlling function—this interaction—was spelt out in Popper's *Spearhead model*, which he proposed as an amendment of Darwinism. The mind grows with the development of language, especially the higher argumentative function. The mind is the producer of human language. But language evolved first. Hence rationality, self-critical thinking, is due to the interaction with World 3. (*Autobiography* 1974: §§39, 40)

Perhaps Popper was wise to regard this interactionism simply as a conceivable solution, one which could provide 'partial understanding' (*Self/Brain* 1977: 38), just as he thought of Darwinism as an explanation 'in principle'. A complete understanding of the mind-body problem may escape us, but any theory, even an 'absurd one', is better than none at all. (*Parmenides* 1998: 91, 51–2, 246; *C&R* 1963: 141)

Why is this no more than a 'conceivable solution'? On the one hand, there is an obvious lack of rigour in the terminology employed (consciousness, mind, self). More controversially, this type of interactionism requires the 'openness of World 1 towards Worlds 2 and 3'. (*Self/Brain* 1977: 171; E7, §52; Dialogue X) This is a very controversial assumption since the physical world is usually seen as 'causally closed'. As some of the critics were quick to point out the closure of the physical world is a problem for Interactionism. (Feigl/Mehl 1974: 528–29) The causal closure of World 1 would prevent the type of interaction that Eccles and Popper envisaged. If the physical world is indeed closed to immaterial influences—like thoughts—then World 2 cannot interfere in World 1. World 1—the physical world—is only open to physical influences. Popper and Eccles casually observed that their interactionism may require a re-interpretation of the First Law of Thermodynamics as a statistical law. (Dialogue X) The First law states that the energy of the universe, regarded as a closed system, is constant. Energy is never lost but it is transformed from useful to useless energy. Interactionism, however, requires that 'mental energy' impacts on 'physical energy'.

Popper's and Eccles' stipulation of the causal openness of World 1 is simply a redescription of our familiar experience that mental events can affect physical events. Even if the First law is re-interpreted as a statistical law it does not solve the mystery of how the mind can impact on the body. This is the great puzzle people have tried to solve since Descartes's dualistic hypothesis.

Popper in fact conceded that the 'emergence of consciousness is a miracle or mystery'. (*Open Universe* 1982/1988: 150) Interactionism solves this mystery as little as the rival approaches. *The Self and Its Brain* was in part an extended restatement of the mind-body problem. The view that the mind emerges from the brain and takes on a life of its own complies with our intuitions. Materialism has difficulties explaining how the mind is reducible to neural events. But the term 'emergence' is deceptive when applied to the mind-body relationship. Consider my previous example. It is true that the cake can no longer be dissolved into its ingredients. But the emergence of the cake is not a mystery. The cake is as material as its ingredients. The heat transforms the dough into a cake. Heat is just molecular energy. The mystery of the mind-body problem is how the *immaterial* mind and its mental processes can emerge from the *material* brain and its neural processes.

This mystery still puzzles researchers. The problem is, as the neuroscientist Baroness Susan Greenfield remarked, how to cross the Rubicon from the physical to the mental. Neuroscience must explain how to 'convert' the water of objective brain events into the 'wine' of subjective consciousness. A tension exists, however, between these two perspectives, between what we feel subjectively and what neuroscience can explain objectively. Neuroscience should explain how the subjective mind works. So there must be in the brain a correlate to the subjective mind. For Greenfield and her Oxford team the correlate exists in neuronal assemblies, which she defines as

> …variable, highly transient (sub-second), macro-scale groups of brain cells (for example, about 10 million or more) that are not confined to, or defined by, anatomical brain regions or systems. (Greenfield 2016: 43)

The neuronal assemblies exist at a mesoscale, midway between the micro- and the macro-level. In this scheme the neuronal assemblies do all the work. But how do they do it? There is no claim that they causally explain how the qualia—instances of subjective, conscious experience—come about. Furthermore, this neuronal approach denies that there is a one-to-one mapping between particular brain states and mind states. The brain does not have a 'centre' which directs operations. Rather, brain regions contribute to the whole. In the language of neuroscientists, the brain is 'holistic'; i.e. several regions of the brain are involved in mental activities. The brain is also adaptive and responsive to changes in the environment. Colour, for instance, is employed to influence consumer behaviour: red to sell luxury goods. It has also been linked with various mental perceptions. Red is associated with warmth and cosiness, blue is associated with creativity and imagination.

The time-honoured Cartesian problem even plagues the notion of emergence: For how can the neuronal assemblies, which are physical in nature, bring about the immaterial mental qualia? Greenfield admits that even emergence does not deliver a causal explanation—it seems it is correlations all the way down. (Telephone Interview 22/06/2017)

Popper, of course, was not very interested in qualia, the subjective side of World 2. His interest resided in how the subjective mind of World 2 can interact with World 1 and produce the objective thoughts, which will find a home in World 3. But Interactionism fails to provide a satisfactory explanation. Popper's initial scepticism regarding the mind-body problem was justified.

9.3.2   Evolutionary Epistemology

Popper's evolutionism aimed to cover both the growth of knowledge and the emergence of mind in a physical world. A further consequence of his evolutionary turn was his adoption of what is known as evolutionary epistemology. It is perhaps not surprising that Popper also came to endorse an evolutionary epistemology—a theory of knowledge based on evolutionary considerations. It was not only inspired by Darwin but owed much to Kant. Such an evolutionary view had predecessors in thinkers like H. Spencer, G. Simmel, C. S. Peirce and James M. Baldwin, who lent his name to the Baldwin effect. (See Campbell 1974: Chap. 12.4) The main proponents today, apart from Karl Popper, are Donald T. Campbell and Popper's lifelong friend Konrad Lorenz. Popper had known Lorenz (and his future wife) since 1908–09, 'the time before the automobile'. Like his father, the young Lorenz was an 'extraordinary personality' [321.04; my translation]. The two men lost sight of each other until Lorenz discovered Popper's epistemology through Donald Campbell, as he wrote in a letter (22/10/69 [321.04]). From then on, they renewed their friendship and visited each other. Lorenz, who shared the 1973 Nobel Prize in Physiology or Medicine (with Nicolaas Tinbergen and Karl von Frisch) is often regarded as one of the founding fathers of the study of animal behaviour (ethology). Yet in his work, he makes frequent references to Kant.[11] How, then, do you 'marry' Darwin and Kant? Kant conceived his famous categories of thought—space, time, causality, etc.—as a priori given. They are the lenses through which the world is comprehended. They are rigid and necessary presuppositions of our ability to acquire objective knowledge about the external world. Campbell, Lorenz and Popper accepted the existence of a pre-given framework, the priority of theory over experience. Already in Grundprobleme (1979: 76; 2009: Chap. IV, §9) Popper clearly agreed with Kant that experience without a priori presuppositions was impossible. Yet all three denied the

---

[11] See for instance his book *Behind the Mirror: A Search for a Natural History of Human Knowledge*, which first appeared in German in 1973.

absolute validity and rigidity of Kant 's categories of thought. They must be thought of as evolutionary adaptations. That means that they are a priori only from an individual organism's point of view but a posteriori from the point of view of the history of a species. They are fallible.

Konrad Lorenz (1903–1989), Popper's lifelong friend. *Source* Wikimedia Commons

In a lecture given at the *LSE* in 1989, Popper proposed that the 'origin and evolution of knowledge' was as old as the 'origin and evolution' of life. He called knowledge a priori if an organism possessed it *prior to sense experience*. It was *inborn* knowledge. A posteriori knowledge was acquired through the use of an organism's senses, in response to environmental inputs. (*Propensities* 1990: 46; italics in original) Popper's characterization is in line with the standard philosophical usage of this terminology.

In his *Logic* (1959/1980: XV, 15) Popper asserted that the 'central problem of epistemology' is 'the problem of the growth of knowledge'. He also denied the existence of pure observation statements. 'For our senses to tell us anything, we must have *prior* knowledge'. (*Propensities* 1990: 37; italics in original) But the first clear indication of his endorsement of evolutionary epistemology occurred in a lecture given at Cambridge in 1953 (reprinted in *Conjectures and Refutations* 1963: Chap. I). He started out by dismissing as absurd the seventeenth-century rationalist view of inborn ideas. By contrast, 'every organism has inborn *reactions* or *responses*'. One may even speak of inborn (unconscious) knowledge.

> This 'knowledge' is not, however, *valid a priori* [as Kant had assumed]; an inborn expectation, no matter how strong and specific, may be mistaken.
>
> Thus we are born with expectations; with 'knowledge' which, although not *valid a priori*, is *psychologically or genetically a priori*, i.e. prior to all observational experience. (*C&R* 1963: 47; italics in original)

There is, for instance, the expectation, analogous to Kant's 'law of causality', to find regularities in the world. The 'Kantian categories', however, are not fixed but shaped by evolution. They are mental adaptations. In this sense the Kantian a priori appears as the a posteriori of our species history. (*Offene Gesellschaft* [3] 1983: 65;

Zukunft [4]1990: 29–31; *Propensities* 1990: 45–7) Kant was right that humans operate with a mental scaffolding but it evolved as the result of adaptations. As Darwin had already stressed, these adaptations are never perfect. Hence a priori knowledge becomes, in Popper's hands, conjectural 'knowledge'. He went further than Kant and claimed that '99% of the knowledge of all organisms is inborn and incorporated in our biochemical constitution'. (*Propensities* 1990: 46) Popper also went beyond Lorenz. (Cf. Munz 1993; Munz 2006) True, evolution has moulded our cognitive apparatus. But there never has been observation without prior a priori knowledge. Popper offered an updated version of Kant's apriorism. A priori knowledge is neither valid nor necessarily true. It is hypothetical. Popper's apriorism is entirely conjectural. It is also more radical than Kant's. For Popper defends the thesis that

> ...all knowledge is a priori, genetically a priori, in its content. For all knowledge is hypothet-
> ical or conjectural: it is our *hypothesis*. Only the *elimination* of hypotheses is a posteriori, the
> clash between hypotheses and reality. (*Alles Leben* 1996: 127–30; *All Life* 1999: Chap. 4,
> pp. 45–47; italics in original)[12]

All life is problem-solving so that new forms of the a priori arise through inventions and 'mutations' (a reference to his *Spearhead model*). These new theories and expectations are always prior to observations. They are not yet built into our 'physiology'. They are our inventions. 'The a priori is creativity.' (*Offene Gesellschaft* [3]1983: 70; *Zukunft* [4]1990: 30–31)

If the origin of knowledge coincides with the origin of life, then the evolution of knowledge becomes a problem of cosmology. This link drew Popper's attention to a new theory of the origin of life. Its author was the afore-mentioned Günter Wächtershäuser who continued his scientific work even after having become a patent lawyer. His theory of the origin of life only assumed the existence of 'chemical reactions confined to monomolecular layers of organic constituents...'. It replaced, according to Wächtershäuser, the so-called soup theory or 'broth theory' of the origin of life, which required amino acids as organic building blocks for life.

> It is here proposed, at an early stage of evolution, there are organisms drastically different
> from anything we know. These organisms possess neither cells nor a mechanism for division
> and yet they possess growth; they possess neither enzymes nor a mechanism for translation,
> yet they possess an autocataclytic metabolism: they possess neither nucleic acids nor any
> other template, yet they still possess inheritance and selection. Although the meagreness of
> their possessions are barely enough to call them living, they do add up to a capacity for
> evolution (16/06/86) [566.22].

This new theory satisfied one of Popper's methodological rules in that it had more explanatory power than its predecessors.

Wächtershäuser also produced a biochemical theory of the evolution of the eye. The invention of the eye solved a problem of adaptation. Popper agreed. (*Propensities*

---

[12] In an unpublished handwritten note on evolutionary epistemology, Popper underlined that 'every-thing is genetically a priori' [247:13, my translation]. In a letter to Lorenz (21/01/75) he maintained that there is a 'historical-genetic-biological apriori' but no 'apriori (=absolutely certain validity)' [321.04; my translation].

1990: 40–5) He linked Wächtershäuser's theory to the role of a priori conjectural knowledge.

> The invention of the eye is (…) an invention of new theoretical a priori knowledge, of an adaptation to the environment….It is theoretical knowledge of a high degree of universality, almost like Kantian knowledge of space and time. (*Propensities* 1990: 49; italics in original; cf. Miller 2006)

Wächtershäuser repaid the interest by endorsing Popper's theory of science. Popper's greatest achievement was the destruction of the philosophy of inductivism, reductivism and determinism. He deserved praise for emphasizing the role of eliminations of tentative solutions, of failed hypotheses. (Wächtershäuser 1995; [566.22])

Popper's struggles with the mind-body problem show that after his retirement he continued to be as active and dynamic as ever before. He remained so until his death in 1994.

Chapter 10
Life After Retirement

For I believe in the power of ideas, including the power of false and pernicious ideas. And I believe in what I might call the war of ideas. (K. Popper, *Conjectures and Refutations* 1963: 373)

On April 26, 1963, the *LSE* had asked Popper, who was approaching the age of 62, whether it was his 'desire that the School should ask the University to reappoint you to the higher retiring age, i.e. to the end of the session 1968–69'. Popper was pleased (05/05/63) as he had turned down a lucrative offer 'from one of the Eastern Universities' [549.04]. When Popper retired from the *LSE* in 1969, he was world famous. Yet financial woes continued to plague him. He retired on a pension of approximately £500 a year, plus a lump sum of £12,000 [334.03].[1] Von Hayek tried to persuade him to come to Salzburg but Popper was undecided. He felt that anti-semitism was still rife in Austria, that his pension was very poor and Hennie would be left without money. Also his influence in Austria was 'nil' compared to Germany. He confessed that he was not popular with most of his colleagues [305.15]. Yet he was in demand as a political and social commentator. Mrs. Melitta Mew, Popper's full-time secretary from 1985, told Simkin (18/05/92) that they declined 90% of requests for media interviews; even the remaining ones were stressful [563.01].

10.1 A Grandee

Upon his retirement in 1969 Popper became Emeritus Professor of Logic and Scientific Method at the University of London. A very special honour followed. Prompted by a suggestion from his old friend at the Royal Society, Peter L. Krohn (an endocrinologist), Medawar decided to nominate him for a Fellowship at the Royal Society

---

[1] Yet he complained to Schilpp (29/03/71) that apart from the one-off payment of £12,000 he would have to live on a monthly pension of £30 [334.03].

© Springer Nature Switzerland AG 2022
F. Weinert, *Karl Popper*, Springer Biographies,
https://doi.org/10.1007/978-3-031-15424-9_10

of London (27/05/71). In the course of their correspondence, Popper admitted to Medawar (10/07/71), in confidence, that the

> …Fellowship of the British Academy is unfortunately an altogether unlucky affair, at least as far as the Philosophy group is concerned. They admit masses of people who are no good…..

Popper's Bronze Bust is housed in a downstairs office of the Lakatos Building, photographed by the author (04/10/2018)

After having failed twice to get 'an old Philosophy Professor from Vienna' appointed as a Corresponding Fellow, he concluded that 'the British Academy is a clique, and I am not proud to belong to it' [325.27]. But he was keen to become a Fellow of the Royal Society although he did 'not count on a favourable outcome'.[2] He needed at least six signatories to support his nomination. Medawar assured him (02/07/71) that he had his support and that he could count on Hermann Bondi, Jack Eccles, Peter Krohn, Solly Zuckerman (the British public servant and zoologist) and perhaps the physicist George Thomson. For the nomination the Royal Society also needed assurance that Popper had British nationality. Popper confirmed (23/07/71) that he had become a 'naturalized British subject on the 1st day of October 1945, in New Zealand' (…) and 'became registered as a Citizen of the United Kingdom and Colonies on June 17th, 1949 (…)' [325.27]. But it was not until June 17, 1976, that Popper was elected. Medawar revealed that he also had the support of the biologists Francis Crick (Nobel Prize, 1962), Jacques Monod (Nobel Prize, 1965) and the French microbiologist André Lwoff (rather than George Thomson). Even before the official induction took place on November 4, 1976, Popper received hundreds of letters of congratulations [405.09]. The election, which was widely reported in the

---

[2] The Popper Archives contain a letter from Medawar to Popper (dated 16/03), in which he expressed his dismay that the Royal Society did not 'have the sense to elect you into the Fellowship' [325.25]. As the letter gives no year date, it must have been written in 1972 because Medawar had earlier warned Popper (07/07/71) that the election would not take place until March 1972 [325.28].

press, required Popper to sign the Charter Book, which featured the signatures of Maxwell, Newton and Einstein [325.28]. Popper did not immediately realize that he had been elected under Statute 12, which was reserved for people who were honoured for their 'conspicuous service to the cause of science (...)'; under Statutes 1-10 working scientists were elected. He thus found himself in the company of Margaret Thatcher and David Attenborough, who became Fellows in 1983. It led to accusations in *The Guardian* (19/10/83) that the election had disregarded proper procedures. The Fellows of the Royal Society were thereafter called upon to reconsider Statute 12: should it be amended or abolished? In May 1985 the President of the Royal Society called a special meeting whose business was, amongst other items, a consideration of the Council's proposed rewording of Statute 12. As a Fellow, Popper received all communications. In a letter (20/05/85) to the President, Sir Andrew Huxley, he expressed his dismay that on the Fellowship List his name (and that of others elected under Statute 12) bore a 'discriminating asterisk' [345.06]. He feared that he would lose privileges, such as submitting a paper to 'one of the journals of the Royal Society'. Huxley (29/05/85) assured him that he enjoyed the same rights as all other Fellows. He should regard the asterisk as a batch of honour since only one person per year was elected under Statute 12, but 40 under the ordinary Statutes. A year earlier (18/07/84) Medawar had already tried to placate him, telling him that he should not feel diminished, since the name of Winston Churchill, whom Popper admired, also bore an asterisk in the Year Book [325.29].

Other philosophers had become Fellows of the Royal Society, for instance John Locke and Bertrand Russell. But 'Popper is the only philosopher in modern times to have been elected a Fellow of the Royal Society primarily in recognition of his philosophical achievement'. (Miller 1997: 369) Many honours had already come his way. For instance, in 1973 (May 25) he received the Danish Sonning Prize in Copenhagen, of which Russell had been the recipient in 1960.[3] In 1976 the Lippincott Award of the American Political Science Association honoured him for *The Open Society*; the Republic of Austria gave him the Grand Decoration of Honour in Gold ('Großes Goldenes Ehrenzeichen'). A reference letter in support of Alan Musgrave (17/02/76) was signed: 'Karl Popper, F.B.A., Correspondant de l' Institut de France, Professor Emeritus in the University of London' [553.13]. The Lippincott award was followed, in 1984, by the Tocqueville Prize for political literature. His books were translated into some forty languages, including Mongolian. (Brudny 2002: 214; Watkins 1997b: 681; Miller 1997: 403–4)

---

[3] Popper confessed to Lorenz (18/02/73) that he had never heard of this cultural prize before [321.04]. It is awarded for outstanding contributions to European culture. He felt that he did not deserve it and that his contribution to European culture was 'non-existent' (letter to Feigl 17/02/73 [536.10]). The award winner received £12,000, as Popper told Magee (31/10/73). In this letter he also proposed some corrections to an article, which Magee published on the occasion of Popper's award of the Sonning Prize [322.18].

The Finnish Society for Critical Rationalism, devoted to the promulgation of Popper's philosophy, invited both Lakatos and Watkins to give public lectures (21/06/73) [360.01]. The French translation of *Logic of Scientific Discovery* finally appeared in 1973. When Popper heard from Ernst Gombrich that Jacques Monod had referred to him in a lecture in London in 1969, he sent Monod a copy of *Logic* (20/09/70) and asked him whether he would consider writing a Preface for the French translation. Monod obliged and wrote a competent and laudatory Introduction. He blamed the belated translation of the book on the obsession of French philosophers with obscure metaphysics [329.23]. He stressed that Popper's *Logic* was one of the few philosophy books, from which scientists could learn lessons about their trade. Monod turned out to be a keen supporter of Popper's epistemology, as there existed broad agreement between their scientific outlooks. He visited Popper in Penn and invited him to give lectures at the Collège de France. He was also instrumental in getting *The Open Society* published in France; translated by his brother Philippe Monod and Jacqueline Bernard [329.23]. It was only in 1974 that French philosopher Jacques Bouveresse published the first important article about Popper's philosophy of science in France. (Brudny 2002: 219)

If all life is problem-solving, Popper experienced his own ups and downs. He was seventy-four and had become hard of hearing. He bought a hearing aid but complained that it started whistling loudly at odd moments. 'It is doubtful', he wrote to Medawar (05/07/77) 'whether I shall keep the hearing aid' [325.28]. He wanted to devote his time to the completion of his Autobiography, as well as his 'Replies' to the essays, which Paul Arthur Schilpp, an immigrant from Germany, was editing for the Popper volume (XIV) of the Library of Living Philosophers. Popper joked in a letter to Magee (February 10, 1969) that someone had aptly renamed Schilpp's Library of Living Philosophers that 'of Moribund Philosophers' [322.18]. The two men had known each other since 1953 when Schilpp invited him to contribute an essay to the Carnap volume he was editing. After Popper's visit to Northwestern University, where Schilpp was then professor of philosophy, he received the official offer (08/05/63) to become the subject of a future Popper volume. Popper accepted the invitation as a great honour, and Schilpp began to send out invitations to potential contributors. Schilpp was disappointed that he had never received so many refusals (04/09/63) but it prompted an intensive search for contributors [334.02]. Popper, naturally, had his favourites. But the mathematician Kurt Gödel, quantum physicist Werner Heisenberg and philosopher Carl Hempel declined the invitation; Isaiah Berlin and David Bohm, although invited, did not contribute. Nor did Solly Zuckerman, who supported Popper's nomination for a Fellowship at the Royal Society. Hanson eventually pulled out, due to overcommitments. Quine at first declined but then relented. Medawar got impatient and withdrew his paper but Schilpp, exceptionally, republished a chapter from *The Art of the Soluble* (1971). Eccles threatened to go back on his promise but changed his mind. Several times Popper urged Schilpp to invite Lakatos, who eventually contributed an essay, which Popper disliked. Kuhn and Agassi also contributed papers but Popper did not want a contribution from Feyerabend. It had taken Schilpp ten years to produce the Carnap tome. The Popper volume also dragged on, much to Schilpp's annoyance. Popper started work on his Autobiography as early as 1965 but Schilpp had to put a lot of pressure on him to finish it. Popper complained that progress was slow and 'I find the struggle with the English language more difficult and time consuming than ever before'. He appealed to Magee for help (10/02/69): 'I need somebody to point out to me all those places where the English is heavy or wrong and to correct so far as it is possible the many mistakes I must have made' [322.18]. Magee was ready 'to start whenever you are' (12/02/69), and so copies were sent to him [322.18].

It did not help that Popper, in order to supplement his pension, had accepted an invitation to teach a semester at Brandeis University in the autumn of 1969, leaving London on August 26. (Brandeis is a private research university in Boston, Massachusetts.) He found the students charming but he did not consider it a pleasant experience [305.05; 305.15]. He shared with Feigl what was wrong (08/10/69): he had only a 'nodding-on-the-corridor acquaintance with my colleagues here, and even this only with two of them'. Besides, they lived twelve miles from the university campus and six miles from the nearest shop, without a car [294.06]. When Yourgrau asked 'My dearest Karlissimus' for a letter of reference for a position at Brandeis, Popper revealed (10/04/68) that he had had a bad experience the previous year. He was interested in a 'tremendous job offer' from Brandeis, but nothing came of it [364.11]. (Yourgrau did not get the job either.) But on the occasion of its centenary, Boston University organized a colloquium in honour of Popper's philosophy of science. Popper finished at least the first draft of his Autobiography in 1969, much to Schilpp's relief. But then Schilpp was getting worried about the Replies. Popper struggled with his Replies to the thirty-three essays on his philosophy. By June 1971 he had only completed a few. Schilpp, who once addressed Popper with 'Mein lieber Karolus Magnus' (18/07/74), confronted him with a plain choice: either he would propose a definite date for the Replies or the project would be shelved in favour of Volume XV, devoted to the French existentialist philosopher, playwright and music critic Gabriel Marcel [334.04]. Enlisting the help of David Miller and Jeremy Shearmur, who read the contributions, and a small team of assistants and typists, Popper took six weeks to comply with Schilpp's demands. Schilpp received the Replies and the revised Autobiography in August 1971. As late as May 1972, he had to admonish Popper to stop sending further corrections [334.03]. (Cf. [328.01]; [330.15]); Miller 1997: 397)

Popper had such great esteem for Schilpp that he nominated him for a Fulbright Visiting Professorship at the LSE, which, however, never came to fruition [334.02]. But Popper, who told Schilpp that he admired John and Robert Kennedy (02/05/75), shared with him the proceeds from his Fontana edition of the Autobiography. He transferred £283.71 or $498.09 into his account [334.04]. He doubted, however, that his Autobiography was 'any good' (08/11/73). This 'Schilpp-Marathon' had put a terrible strain on Karl and Hennie. Besides, he was upset that by the middle of 1976 his Schilpp volume was practically unavailable in the UK so that hardly any reviews had been published in British magazines, and few in the US [334.04; 323.01].

Meanwhile Popper did not neglect his family obligations. He and Hennie visited his sister Anna in Ascona, October 1971 [349.13]. Two years later he expressed his concern for Annie's health in letters to his friend Konrad Lorenz and planned to visit her again [321.04]. Hennie herself underwent a radical mastectomy operation in 1976 [364.12].

10.2   The LSE and Beyond

Popper's relations with his old department in the early 1970s were not cordial. He complained about the behaviour of his former colleagues towards him in a letter to Schilpp (18/04/71): they had not kept their promise that he would be able to continue until he was seventy [334.03]. He gave a paper at the *LSE* but was disappointed by the reception (letter to Jarvie 29/01/75). It 'was such that I decided "never again"' [313.06]. He lamented that all his friends, except Ernst Gombrich and Paul Osters, had deserted him (letter to Schilpp 29/03/71 [334.03]).[4] Magee organized a celebration on the occasion of Popper's 70th birthday (1972). It cheered him up after all the 'depressing things which happened to me since my retirement' (29/07/72). Yet he felt that Magee overrated his 'significance as a philosopher, or my achievement' [322.19]. Still, various grants paid by Leverhulme, the Rockefeller and Nuffield Foundations allowed him to retain an office at the *LSE*. The money also paid for a research assistant—Jeremy Shearmur—whose main task was the procurement of books and periodicals delivered to *Fallowfield*. It often enabled Popper to restrict 'his attendance at the L.S.E. to one day each week'. (Miller 1997: 386) When Popper stopped applying for the renewal of his research assistants, in 1985, Mrs. Mew took over as his full-time secretary and Popper asked for her salary to be increased. Yet Popper felt that he had lost influence. On June 23, 1982, he was awarded an honorary degree at Oxford University. He considered that without Magee's (and philosopher Anthony Quinton's) support he 'would have been as much a dead horse in Oxford as in London' (letter of 26/04/82 [323.04]). Popper was dismayed, as he wrote to Soros, that 'quite a few of my pupils have not only become enemies of my philosophy but silly and arrogant'. So he was buoyed by the news (03/09/87) that Soros had sponsored a conference about his philosophy at Wuhan University in China [563.15]. Soros had done more: he established a Karl Popper Stiftung in Switzerland (1989), a Karl Popper Fellowship at Cape Town University and the *Open Society Fund* offered Karl Popper Bursaries. Popper admired his achievements [351.16; 563.15]. He was also pleased by Watkins's suggestion (16/07/74) to introduce some younger colleagues

---

[4] Osters's name appears in the Foreword of the German translation of *Unended Quest.* Popper thanked him for his help with proof-reading. He is neither mentioned in the *Autobiography* (1974) nor in *Unended Quest* (1976).

to him in an effort to overcome 'the gap which (...) opened up after your retirement' [360.01].

He was not a 'dead horse'. In 1976 he received an honorary degree from City University [232.01] and Salford University [405.09]. Two years later, the city of Vienna awarded him the Karl Renner Prize (1978). Yourgrau tried to nominate him for a Nobel Prize in Literature [364.12; 568.06]. It is worth mentioning that the *London School of Economics* has no building named in honour of Karl Popper. But it runs a Popper Seminar; it hosts a Sir Karl Popper Memorial lecture series, which is held, in principle, every two years; it is funded from the Sir Karl Popper Memorial Fund, which was established in 1994, the year of Popper's death. It offers Popper Prizes for the best performing students. It also has a much-viewed bronze bust, on display in its Departmental Office on the ground floor. It was presented by the Austrian ambassador at the time of Popper's centenary. Another bust is at the University of Vienna. There is still much to remind visitors of Popper's career at the *LSE*. 'Your spirit', as John Worrall wrote to Popper (29/10/77), 'still looms large in our department' [363.24]. Money was the reason why the building formerly known as 'Tymes Court' was renamed after Lakatos, as Worrall explained to me. It now houses both the Department of Philosophy and the Centre for Philosophy of Natural and Social Science. This building was given to the School by a donor who requested that it be named after Lakatos, not Popper. The donor held Lakatos in higher esteem than Popper.

The 1980s started with the news that Italian television was preparing a documentary on Popper's work. It would feature interviews with Helmut Schmidt and Bryan Magee [323.03]. Cambridge University honoured him with the Degree of Doctor of Letters on June 5, 1980 [325.29]. On January 15, 1985, the *New York Times* ran a piece on the scientific status of Freudianism. Grünbaum had argued that at least some psychoanalytic hypotheses were testable. Popper was asked to respond (10/01/85). He did so in a handwritten telemessage: 'any conceivable human behaviour can be explained in the light of Freud's theory'; it made no testable predictions [331.35; cf. *C&R* 1963: 35]. A year later, Popper conferences were held at Edinburgh and Coventry Polytechnic [549.19]. Magee heard from the Oxford philosopher William Newton-Smith that Popper enjoyed 'immense eminence in China'; all his works had been translated and *Logic* had 'sold over 100,000 copies' (May 2, 1988) [549.19; 550.01].

Popper had worked on the galley proofs for his *Postscript* until 1963 but it was not until the autumn of 1969, when Arne Petersen became his research assistant, that work could continue. By that time Popper was busy writing his Replies to the Schilpp volume essays so that the *Postscript* was neglected again [322.16]. Around 1980 additional material was added. What had been planned as a single book became the three volumes of the *Postscript* which were finally published in 1982/83. At first Tom Settle was supposed to edit these volumes. But when he failed to devote the necessary energy and time to the task, Popper entrusted W. W. Bartley with the editorship. Bartley went to great lengths, including the search for suitable publishers. Popper was so grateful that he proposed to share the proceeds from the books with him. They finally agreed that Bartley would receive royalties after Popper's death [272.09]. Their friendship had been rekindled to such an extent that Bartley proposed, in 1981, to write Popper's biography [273.01] Shortly afterwards, Popper heard from Bartley (05/02/83) that von Hayek had also asked him to write his biography [273.03].

In 1983 Bartley got involved in a dispute with an author from the University of East Anglia, whose interpretation and representation of Popper's fallibilism he had dismissed as 'incompetent'. The author threatened to sue Bartley for what he deemed were 'libellous remarks'. Naturally, Bartley turned to Popper for support. There is a draft letter in Popper's correspondence to an editor at Harvester Press in defence of Bartley. It provides an insight into Popper's personality. He writes:

> As you may know I am 81 years old and although I have been quite undeservedly successful in my career, that is due to luck and, perhaps, a little to hard work. On the other hand I have throughout my life suffered from severe attacks which if they were personal or libellous attacks (as they frequently were), I usually did not read. But I learned of sufficiently many such attacks to be often hurt.

The unsent letter continues:

> I can assure you that I found your attacks (though they do not seem to me libellous) very hurting and extremely unjust and unfair. (…) I do not think that Bartley's criticism of you is as offensive as your criticism of me; nor does he misrepresent you as you do misrepresent me.

Popper ended his letter (07/07/83) by urging the author not to pursue litigation [273.03]. Bartley eventually agreed to drop the offensive words. Watkins, who had witnessed this affair, did not want to deal with Bartley again. But then Watkins himself got involved in an altercation with Popper. It concerned a certain phrase in Watkins's book *Science and Scepticism* (1984), which Popper found offensive. Some friction had erupted between the two men before. In the early 1960s Popper had accused him of misrepresenting his views on the falsifiability of whole systems and their individual components. Watkins's apparent departures from the letter, if not the spirit, of *Logic* led to charges of plagiarism. Watkins defended himself (05/01-06/01/62) but reproached Popper with a tendency to throw out 'plagiarism accusations' [359.34]. Just after Popper's retirement from the *LSE* Watkins feared that a 'certain estrangement' had arisen between them, partly because Watkins admitted that he behaved badly towards Popper and partly because he felt hurt that Popper often interrupted him (18/06/70; 16/01/70) [360.01]. Their friendship survived but in 1982 Popper accused Watkins of misrepresenting his theory of the empirical basis. In particular Popper took exception to a certain sentence in *Science and Scepticism* (1984: 252). Watkins complained (10/10/82) that Popper shouted at him over the phone, calling the sentence a 'lie' before putting down the receiver. Eventually Popper apologized for calling the offending sentence a lie (28/10/82). But he would not give up and their discussion rumbled on. Watkins saw nothing wrong with the statement, which did little to assuage Popper's anger. He welcomed criticism (04/01/83) but

> I must again protest against the manner in which you treat a serious theory of your teacher. I find the manner personally insulting, and bound to create confusion. (…) I believe that as your teacher, to whom you clearly owe something, I have the right to expect to be treated with a certain amount of respect [360.03].

Watkins protested that his criticism was not a personal insult (12/01/83). The affair did not dampen their friendship: two years later Watkins sent Popper a copy of his book and offered to apply for the renewal of the 'Wolfson and Leverhulme grants' (07/01/85) [360.03].

And then occurred the saddest of events. On Sunday morning November 17, 1985, his companion Hennie, now Lady Popper, died after a prolonged illness. 'She had a terrible death', Popper wrote to Soros (10/09/87) [563.15]. In November 1978 Popper had taken her to Georgetown University Hospital in Washington [306.02]. The treatment for cancer (carcinoma) had gone well but a year later she was again 'very sick' [309.13]. She had chosen to die in Vienna, and so they travelled to their place of birth. She is buried at the *Lainzer* Friedhof, where he would also be laid to rest.

Popper sold *Fallowfield* and moved to Kenley, south of Croydon, in the southern outskirts of London. He continued to work, made public appearances and expressed his views on politics. Near his new house there was a glider club. Melitta Mew told John Watkins that

> …he was fascinated by the takings off and landings, and persuaded someone there to take him up. After that he would have liked to take lessons. This did not happen, but he did go up a second time at the suggestion of a German television company. (Watkins 1997b: 681)

Back on the ground he gave a lecture in Bad Homburg (November 5, 1991) on 'All Life is Problem Solving' before about 20 bosses of the German car industry (from Mercedes to Volkswagen), as he informed David Miller (17/12/91) [552.03].[5] In 1991–2 the *LSE* was in touch again for various reasons. First they thanked Popper for his donation of the microfilms of his papers to the British Library of Political and Economic Science. Then they wanted to add Popper's publications to the School's research output for the Research Assessment Exercise of 1992. And his old department wished to organize a celebratory event on the occasion of Popper's 90th birthday. Popper willingly accepted the invitation to a 'small luncheon party at the School with time for leisured discussion' [549.04]. Shortly before his death, he became the first recipient of the *Open Society* Award (12/06/94) [563.15].

---

[5] Reprinted in *Alles Leben* (1991): Chap. 12/*All Life* (1999): Chap. 9.

Popper's House in Kenley. University of Klagenfurt/Karl Popper Library, Supplements 29;2;2.
Copyright unknown

On September 10, 1994, Popper was admitted to hospital. Doctors diagnosed
cancer at an advanced stage. He died on September 17, 1994. (Miller 1997: 402;
Brudny 2002: 241–2) He was cremated and laid to rest next to his wife, Hennie, at
the *Lainzer* Friedhof. He had returned home.

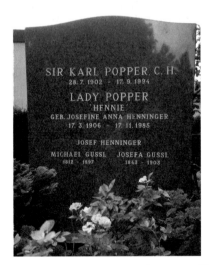

The gravestone of Karl and Hennie Popper, *Lainzer* Cementary, Vienna. *Source* Wikimedia
Commons

10.3 Events and Miscellaneous Thoughts

After his retirement Popper's philosophical influence in the English-speaking world declined. In an article in the *New Scientist*, Anthony O'Hear observed that there was a 'surprising degree of neglect of his thought by contemporary philosophers'. Yet 'virtually no one has achieved so wide an audience outside philosophy as Popper'. (O'Hear 1985) Outside philosophy people were interested in his Critical Rationalism and his views on social and political matters. He was invited to take part in public debates especially in Continental Europe. He participated in panel discussions and symposia, and he gave interviews to journalists from different countries. Bryan Magee did much to present Popper, the man and his ideas, to an English-speaking audience. He popularized Popper's ideas in radio interviews and TV programmes and wrote a feature article in the *Sunday Times* (27/05/73).[6] In June 1973, he proudly informed his friend that his paperback on Popper had sold 'over 7,000 copies in the first 5 weeks'.[7]

Karl Popper at the *LSE* in the 1980s. *Source* Wikimedia Commons

He seems to have had some success. An elementary, non-mathematical introduction to Einstein's relativity theory describes the attitude to theories and experiments in Popperian terms:

> Finally the most crucial point to make about any theory is that it is only as good as its last experimental test. A theory is always on probation and if at any time the experimental evidence is clearly at variance with a theoretical prediction then it is the theory which must be changed. (…) Experiments can therefore invalidate a theory but can never establish a theory

---

[6] A similar article by the lesser known Popperian, Tyrrell Burgess (a professor at the University of East London) applying Popper's ideas to social policies, appeared in the *Sunday Times Supplement* (26/04/74, p. 15). Magee always kept Popper informed about such publications [322.20].

[7] In October sales had increased to 'eight or nine hundred copies a week' but dropped back to 200–400 a week in December [322.19]. In 1978 sales had reached 100,000 copies [323.02] and the book had been translated into many languages, including Farsi and Japanese. By February 1980 the book had become a bestseller in the Fontana Modern Masters Series [323.01].

as inviolate; a theory to be scientific must always be capable of disproof. (Marks 1972: 30; see also Cox/Forshaw 2012: 14)

The author credits Popper for this general attitude to science. In Kenley Popper received many visitors. He wished to get to know Soros better, as they never had 'got near to each other' (22/10/87). He invited him to stay 'for at least three days: I have 3 guest bedrooms (two with a double bed and one with 2 beds) since I have here in my house from time to time a conference with one or two friends and their wives' [563.15, underlined in original]. World leaders knocked on his door: the former German president Richard von Weizsäcker, the Dalai Lama, the emperor of Japan, the former Prime Minister and President of Portugal, Mário Soares and Vàclav Havel, the first President of the Czech Republic from 1993 to 2003. (Brudny 2002: 233) He had been in contact with Helmut Schmidt for a number of years. Schmidt sent Popper a telegram, in 1977, in which he credited the *Open Society* with having helped Western democracies [346.42]. The two men met personally for the first time in December 1980. Shortly afterwards, Schmidt visited Popper in Kenley (November 1982). Popper was receptive to Schmidt's support. He sent him a handwritten note (31/07/82), in which he expressed his gratitude for what 'you represent as a politician, what you are trying to achieve, and what you have achieved' [346.42; my translation].

He had turned into a public intellectual in the best French tradition: he expressed his views on political and social issues of his day. It is not the case that Popper had confined himself to the philosophy of science before, but his social and political commentaries seemed to take centre stage. For instance, he revisited his 'theory of democracy' in an article in *The Economist* (1988), which expanded on his earlier views on democracy.

Before I turn to his 'theory of democracy' let me focus on some more diverse thoughts throughout his work, which, on reflection, are a consequence of his Critical Rationalism. These many scattered remarks range from anti-authoritarianism, internationalism to paternalism and moral responsibility.

Recall that Popper felt a strong antipathy to 'verbiage'. He suspected that it only served to hide vacuity of thought. It turned him against such philosophers as Fichte, Hegel, Heidegger and members of the Frankfurt School. Traditionally the French intelligentsia are also fond of rhetorical flourishes. This was one of the reasons, as I learned from philosopher Jacques Bouveresse, for the late translation of Popper and Wittgenstein into French. He taught then, in the late 1970s, at the *Sorbonne* and in 1995 became Professor at the *College de France* in Paris. Through his writings Bouveresse did much to introduce Popper, Wittgenstein and analytic philosophy in France. He promoted critical standards of thought.

Popper had the admirable gift of expressing complex ideas in simple language. The German weekly *Die Zeit* published, without Popper's permission, an article (24/09/71) entitled 'Wider die großen Worte' ('Against Big Words') [340.13]. He valued intellectual modesty and clarity of expression. In this vein, as I pointed out, he accused intellectuals of having committed 'great crimes' because they had failed to fight against fanaticism and relativism. (*Auf der Suche* [2]1987/*In Search* 1992: I.1,6; III.13–15; *Alles Leben* 1996: Chap. 11; *Lesson* 1998: Chap. 9) He called on intellectuals to resist all brands of irrationalism and intellectual fashions. As part of this blanket condemnation of what he saw as the failure of intellectuals, he dismissed academic disciplines and schools as irrelevant. We are problem-solvers, not guardians of subject matters. Science, he added, has no authority. One should not believe in experts.[8] The job of philosophers, he held, was to 'swim against the tide' (*Parmenides* 1998: Essay 7, §§1, 2) Intellectuals had a moral responsibility. They should be committed to honesty and the search for truth. They should swear by a generalized Hippocratic Oath. (*Myth* 1994: Chap. 6) He went so far as proposing that Nazism could have been prevented if German intellectuals had accepted intellectual responsibility. (*Zukunft* [4]1990: 103)

Swimming against the tide: this advice captures the spirit of his own life as a professional philosopher. He fought a lone battle against the *Vienna Circle* and was amongst the few who opposed subjectivism in physics. He rejected the scientific pretensions of Marxism and psychoanalysis. If opposing the 'zeitgeist' is the role of intellectuals in general and of philosophers in particular, it may come as no surprise that Popper praised the 'omnipotence of ideas'. (*Open Society* II: 107; *C&R* 1963: 332, 373; *Parmenides* 1998: Essay 7, §§1, 2; *Auf der Suche* [2]1987: II.10) In contrast to Marx's official doctrine of materialism, Popper valued the power of ideas. He had experienced it in his own life. After all he had arrived at his conjectural theory of knowledge through learning of Einstein's advance over Newton. (*Parmenides* 1998: 50) When Eddington's expedition corroborated one of Einstein's predictions, Newton was shown to have been mistaken. Freud's dream analysis provoked Popper's views on demarcation. (*Realism/Aim* 1983/1985: II, §18) His study of Greek cosmology, the history of materialism and of scientific ideas and the influence of the Enlightenment further emphasize, in Max Weber's words, that ideas *channel* social action even if they do not determine it. Popper's lifelong fight for objectivity and Critical Rationalism bear testimony to his belief in the influence of ideas on the development of civilization.

Although Critical Rationalism did not establish itself as an academic school, it has its heirs, such as David Miller and Jeremy Shearmur. Two of his most committed followers in Germany are Hans Albert and Hans-Joachim Niemann. Albert applied Critical Rationalism to social and political theory and defended Popper in his dispute with the Frankfurt School. Popper asked Schilpp to invite him to contribute to his Schilpp volume, but he did not take up the invitation. Niemann is a commentator, translator and editor of Popper's work and a defender of active Darwinism.[9]

---

[8] This theme runs through his work: *C&R* 1963: 67; *Objective Knowledge* 1972: 182 [fn30], 310; *Realism/Aim* 1985: Preface 1956, pp. 5–8, 159; *Ich weiß* [2]1992: 24; letter to Isaiah Berlin, dated 21/03/1959 [276.10].

[9] http://www.opensociety.de/Web1/. I owe the reference to Niemann to David Miller.

Fridtjof Nansen (1861–1930), explorer. *Source* Wikimedia Commons

Sir Karl remained interested in music; he expressed some particular views on art. Generally, he preferred objective to subjective art, classical to modern music. He rejected all forms of expressionism, according to which art must be the expression of emotions or the artist's inner feelings. What matters is the work of art itself. 'The main aim of the true artist', he writes in his *Autobiography* (1974: 48), 'is the perfection of his work'. A work of art is a denizen of World 3. This does not exclude that works of art evoke emotions but this exchange with World 2 is not their main function. A work of art should be appreciated for its own sake, irrespective of the artist's intentions. The issue is 'what problem does it deal with?'

Popper's moral views were shaped in his formative years. His thinking was influenced by the socialist ideas of Joseph Popper-Lynkeus, a distant relative and a member of the *Monists*. They were interested in philosophy, science and the reform of society. His early pacifism was the result of reading Bertha von Suttner and Fridtjof Nansen. As a boy Popper had been deeply shocked by the grinding poverty he witnessed in Vienna. He agreed with his socialist friend, Arthur Arndt, that 'nothing could be more important than to end poverty'. (*Autobiography* 1974: 7) This was the job of good governments. Like Kant, Popper believed in ethical individualism, if not in his categorical imperative. Rather he suggested, as much later Habermas did, that we could speak of the rightness of norms.

> We might (…) introduce, in correspondence to the concept of truth, the concept of the validity or rightness of a norm. (Quoted in Shearmur 1996: 92)

He defended the autonomy, moral freedom and responsibility of individuals in a democratic, cosmopolitan society. All humans are equal as rational beings. But he adopted a pragmatic stance in the social sphere. He shared with Weber the endorsement of ethical responsibility. At the beginning of their relationship, Magee quizzed Popper (12/11/58) about his attitude towards the use of violence on the part of oppressed people. Magee himself felt that the use of violence was justified in tyrannic regimes. Nine days later (21/11/58) he received a satisfactory reply: Popper seemed to agree that citizens who are denied self-governance and freedom are entitled to use violence under certain circumstances [549.19].

Popper, as I have said, saw himself as one of the last heirs of the Enlightenment. But there is one aspect of Enlightenment ideas which he did not share. Popper believed in the importance of traditions, which the Enlightenment philosophers in general impugned. They faced authoritarianism, religious intolerance, censorship, irrationalism and despotic rulers. It is no surprise that they took a dim view of traditions. Not so Popper. He even proposed a 'sociological' or 'rational' theory of traditions. (*C&R* 1963: Chap. 4; cf. Shearmur 1996: Chap. III; Simkin 1993: §21)

According to Popper, a theory of traditions should be sociological in nature. Firstly, the social sciences must explain the unwanted and unintended consequences of social actions. They must understand the functioning of institutions, which themselves may produce unforeseen consequences. They must explain how they arise. Secondly, they must understand the role of traditions: they provide social life with order, predictability and regularity. Again, he saw a similarity with scientific theories. They are 'instruments' which allow us to 'bring some order into the chaos in which we live so as to make it rationally predictable'. (*C&R* 1963: 175) Traditions are intermediaries between persons and institutions. They support the workings of individuals and institutions. But traditions may also help to curtail the malfunctioning of institutions. There are both similarities and differences between institutions and traditions.

> Institutions and traditions have much in common; among other things that they must be analysed by the social sciences in terms of individual persons, their actions, attitudes, beliefs, expectations, and interrelations. (*C&R* 1963: 178)

Here Popper repeats his early commitment to methodological individualism. (The reader may recall that Popper did not reduce sociology to psychology. His methodological individualism includes a situational analysis of institutions.) He characterizes an institution as 'the common observation of a certain set of norms for joint fulfilment of certain *prima facie* social functions, public or private'. Examples of institutions are schools and universities, the police force, the British Academy, the Monarchy and an insurance company. Traditions, by contrast, 'describe a uniformity of people's attitudes, or ways of behaviour, or aims or values, or tastes'. Traditions extend over generations and influence people's behaviour in similar ways. Examples of traditions range widely from religious ceremonies and rituals, differences in lifestyles, to national food preferences and table manners. For Popper the tradition of clear thinking and speaking, of reason, takes pride of place. (*C&R* 1963: 178–9, 181; cf. Simkin 1993: 149)

One important additional function of traditions is that they also provide the material for criticism and change. Criticism always happens against a background of traditions. It is a dogmatic attitude, which is also required in science. Popper repeats his fundamental idea, going back to the *Open Society*, that 'history is not a river'. (*Lesson* 1997: 40–1) We can learn from the past, but we cannot foretell the future. As what the future holds in store cannot be predicted, historicism must be rejected. The consequence is indeterminism. Nevertheless, if ideas exert an influence in history, as

Popper continued to emphasize, it becomes necessary not only to listen but to respect the lessons of the past. There are, however, two attitudes towards past traditions: a critical and an uncritical stance. (*C&R* 1983: Chap. 4) Given that he was irreverent towards the authority of experts and denied the importance of disciplines he obviously rejected the uncritical respect for past traditions. We are what we are thanks to the work of the past, as Hegel put it. We can build on the achievements of the past. But traditions should not be followed blindly wherever they lead us. Popper credited the Greeks (Parmenides, Xenophanes) with having 'invented a critical attitude towards myths'. Western civilization was based on Greek ideas of 'truth, democracy, justice, humanity, brotherhood of men and cosmology'. And it was this tradition, which Popper had been at pains to revive and carry forward. He therefore transformed the critical attitude towards the history of scientific ideas into a general attitude towards disciplines, institutions and societal traditions.

> … we should always remain conscious of the fact that all social criticism, and all social betterment, must refer to a framework of social traditions, of which some are criticized with the help of others, just as all progress in science must proceed within a framework of scientific theories, some of which are criticized in the light of others. (*C&R* 1963: 178)

Unburdened by political correctness, Popper declared Western civilization to be superior, as it is the only one based on science. (*C&R* 1963: 203) It had invented the 'rationalist tradition'. Given his cosmopolitan sympathies this statement should probably be read as 'Western-style' civilization. His assessment was based on his conviction that Western-style democracies provide the best of what open societies can offer. They cultivate a pluralism of ideas, provide individual freedom and protect against the misuse of power. (*Auf der Suche* [2]1987: 87–8; *In Search* 1992: 70–1) As 'all power is dangerous', the individual needs protection. (*Open Society* II: 25–6) Freedom is more important, in Popper's eyes, than economic success. (*Ich Weiß* [2]1992: 97; *Alles Leben* 1996: 276–7; *All Life* 1999: 112–4; *Lesson* 1997: 43–4) Western societies are open, dynamic societies, which stand in sharp contrast to closed societies. Habermas shared Popper's assessment. A Western-style civilization is superior because of its problem-solving capacity. Nevertheless, Popper did not pretend that Western societies were 'perfect'.

10.4   On Democracy

In an article in *The Economist* (1988), Popper proposed a revised theory of democracy. He contrasted *his* account with the classical view. He identified the latter with the misunderstanding that democracy means 'the rule of the people'. Popper agreed that a majority rule should apply to 'dismiss bad rulers'. But he rejected the view that the people are the rulers. The *vox populi* is a myth. 'Owing to its anonymity, public opinion is an *irresponsible form of power* ….' (*C&R* 1963: 349; italics in original) People control the controller, by democratic means. What prevents harm, the abuse of power, Popper said in truly republican style, is the design of institutions. Democracy is not the rule of the people but the rule of law. A democracy cannot be better than the democrats. (*Zukunft* [4]1990: 110) He told Russian readers of *The Open Society*— the translation appeared in 1992—that the main idea of the book was the rule of law. It was his firm belief, which he reiterated in a letter to *The Times* newspaper (14/10/58). 'I believe in an established legal framework and I do not believe in leadership', and certainly not in 'ad hoc decisions of wise leaders' [355.06].[10] Popper followed Kant in holding that only the rule of law guaranteed civil liberties and the equality of citizens.[11] But the rule of law guarantees only freedom and justice, if it is based on the right kind of constitution. Such a constitution must (a) share the burden of citizenship, both advantages and disadvantages, equally amongst citizens; (b) make laws and courts act impartially towards all citizens; and (c) not allow any institution or person to stand above the law. (*Open Society* I: Chap. 6, p. 89; *Alles Leben* 1996: Vorwort; Chap. 9; *All Life* 1999: Preface; Chap. 8) Popper approvingly quotes Kant's description of a just constitution as

> ...a constitution that achieves the greatest possible freedom of human individuals by framing the laws in such a way that the freedom of each can co-exist with the freedom of all others. (*Open Society* I: Chap. 6, p. 247 [fn4], italics in original; cf. Shearmur 1996: 111–12)

As we have seen, Popper also adopted Kant's cosmopolitanism. Kant had raised Locke's social contract idea for individuals in one society to the level of nation states. He argued for the creation of a confederation of states to guarantee peace and prosperity amongst nations. Popper, too, was in favour of peace and international organizations. He referred to the *League of Nations*. (*Lessons* 1997: 52; *Alles Leben* 1996: 325–26; *All Life* 1999: 142–43) But his cosmopolitanism took the form of a rejection of all shapes of nationalism. 'Nationalism is evil' because it creates the false impression that nations are unequal. It lowers the threshold for potential conflict because the citizens of each nation are encouraged to think that their own nation is superior to any other. Their nation should enjoy priority over others. Anti-nationalism is a strand in Popper's thinking which harks back to his early pedagogic publications. His second essay discussed the 'Philosophy of the *Heimat* Idea' (1927) in an educational context. He voiced his opposition to nationalism in *Open Society*:

> The idea that there exist natural units like nations or linguistic or racial groups is entirely fictitious. (…) The principle of the national state... owes its popularity solely to the fact that it appeals to tribal instincts. (Vol. I: Chap. 9, n. 7(1); cf. Hacohen 2000: 437)

And further:

---

[10] This did not stop him, in August 1989, to send Mrs. Thatcher two handwritten letters to thank her for the leadership she had shown in saving 'our country'. 'Now, the only serious sickness is Anti-Thatcherism which is a nasty but silly ideology…' [565.07]. Thatcher's Press Secretary thanked Mr. Potter (14/08/89) for his recent letter. An official invitation to Luncheon at 10 Downing Street in honour of French President François Mitterand on October 25, 1984, was a further sign of Popper's public recognition.

[11] Incidentally, Popper did not believe that all humans were equal in every respect but he did reject the existence of privileged elites; so he wrote to the German philosopher Jürgen Mittelstraβ on December 18, 1984 [329.09].

None of the theories which maintain that a nation is united by common origin, or common language, or a common history, is acceptable, or applicable in practice. The principle of the national state...is a myth. It is irrational, a romantic and Utopian dream. (Vol. II: 49; cf. *Auf der Suche* [2]1987: 131; *In Search* 1992: 120)

This warning of the 'danger of nationalism' runs through his work, for his *Autobiography* (1974: 83) repeats that all 'nationalism or racialism is evil, and Jewish nationalism is no exception'. In order to prevent war Kant had pleaded for a 'social contract' between nations. Both trade links and international treaties would reduce the risk of conflict. Whilst Popper agreed, he abandoned the pacifism of his youth. He was in favour of 'war on war'. That is, he believed in just war and nuclear deterrence. (*Myth* 1994: Chap. 6; *Lesson* 1997: 51; *Auf der Suche* [2]1987: 243–5; *In Search* 1992: 213–5) 'We should not shrink from waging war for peace'. (*Alles Leben* 1996: 288; *All Life* 1999: 119) He was opposed to unilateral nuclear disarmament, as he wrote in *The Times* newspaper (11/04/87) [565.12].

Let us turn from external to internal matters, from foreign to home affairs. As far as internal politics is concerned, Kant was opposed to all forms of 'paternalism' or state intervention in the lives of private citizens. Although Popper regarded all power as dangerous (*Open Society* II: 129), he nevertheless thought that some form of state intervention was necessary in the interest of the common good.

Social life without any kind of "repression" is an old dream. Of the practicable ideals not greatly more repressive than anarchism, the least repressive would involve that minimum amount of state interference compatible with equality in the distribution of the burden of citizenship. ('Replies' 1974: 1163)

For example, Popper was in favour of some form of control of the media in order to protect the young. He accused the media, especially television, of a corrupting influence on young minds. In an interview with the Italian journalist Gioncarlo Bosetti, he went so far as to call for censorship of the media if they failed to exercise more restraint and responsibility. (*Lessons* 1997: Interview II) He was also in favour of interventionist health policies. Alcohol consumption and smoking should be restricted by law. Apart from control of the media, an open society now had other priorities: securing peace and halting the population explosion. Public policy should not aim at maximizing happiness; its job is to 'minimize misery'. (*Myth* 1994: 124; cf. Shearmur 1996: Chap. IV–V; Simkin 1993: Chap. 21) He defended a kind of 'negative utilitarianism': 'Work for the elimination of concrete evils rather than the realization of abstract goods'. (*C&R* 1963: 361) In line with his republican sympathies, Popper preferred indirect state intervention through the framework of laws. The right institutions would protect the individual from an overpowering state, just as the right criteria of theory choice would guarantee the rational growth of knowledge. In a letter to Magee in March 1974, Popper shared his views on the problem of nationalization and taxation. He suggested that 'the state should take a share of 51% of all the shares of all public companies'. And that 'only 40% or 41% of the income should go to the state to start with' [322.20; underlined in original; cf. 349.12].

In newspaper articles and interviews he contrasted his own optimism about Western civilization with the widespread pessimism in the media. In an article in

The Times (11/04/87) he reaffirmed that our 'society is the justest that has ever existed', in part thanks to 'much-maligned science, technology and even industry'. There was much left to do but it was 'eager for reform'.

> Its greatest evil is the band of self-appointed high priests who preach that we are living in Hell. This religion is now almost universally accepted; and it is of course sincerely believed by those high priests of the so-called media.

In April 1992 he gave a wide-ranging interview to the influential German news magazine *Der Spiegel* (reprinted in *Alles Leben* 1996: Chap. 14; *All Life* 1999: Chap. 11). He repeated his criticism of Marxism.

> Our liberal social order is the best and justest there has ever been on earth. It rose through evolution from *the one that Marx knew. (All Life* 1999: 117; *Alles Leben* 1996: 285; italics in original)

Popper, incidentally, was also very critical of the Green Party because of its 'crazy hostility' to technology and science. (*Alles Leben* 1996: 291; *All Life:* 1999: 122; *Ich Weiß* [2] 1992: 31–2; 78–9) Had Popper lived longer he would have seen that the Greens are opposed to dirty, old technologies, not clean new technologies. He blamed environmental disasters on the population explosion. 'Really, only children who are wanted should come into the world'. But the population growth should not be solved by government directives. It should be solved through education. Unsurprisingly Popper put his trust in science.

> Unwanted children are at risk, and I mean morally. People who do not want them should have the means not to have them. The means exist already – I am thinking of the abortion pill. (*All Life* 1999: 122; *Alles Leben* 1996: 291–2)

Popper did not offer a theory of political power or social justice. (He did not discuss Rawls's influential theory of social justice although he referred to it in a letter to von Hayek, dated 28/04/1977 [305.16].) But he was strongly in favour of a two-party system, which he would have liked to see in place in democratic countries. It exists clearly in the United States. There is of course a Liberal Party in the UK but power mostly alternates between Labour and the Conservatives. According to Popper such a system has the advantage of barring smaller parties permanently from gaining access to the keys of power.[12] It produces clear majorities and strong governments, which can be held responsible for their policies. (Apparently it did not disturb Popper that political participation in the UK is based on a 'first past the post' voting system which typically returns parties to power with as little as 35% of the popular vote. In the United States a presidential candidate can win the popular vote, as did Hillary Clinton in 2016, and still lose the elections because most states accord their electoral college votes on the basis of a 'first past the post' arrangement.) Popper felt that

---

[12] This seems to be true of the United States, where power alternates between democrats and republicans. But historically it is not correct in the United Kingdom. The Liberals were in power over four terms under William Gladstone between 1868 and 1894. The party returned to power in 1905 and remained in power until 1915. After a long absence from Downing Street the Liberal Party formed a coalition government with the Conservatives under the leadership of David Cameron in 2010.

proportional representation was 'disastrous', because it was not clear who was in charge. (*Economist* 1988; *Ich Weiß* [2]1992: 10, 65–6; *Alles Leben* 1996: 244–45; *Lesson* 1998: 84–85; [565.12]) After reading a Bernard Levin article in *The Times* he wrote to this most famous journalist of the day (probably in April 1987) to voice his opposition to 'electoral reform in the direction of proportional representation'. The arguments in favour of it were based on a 'mistaken theory of democracy = rule of the people: since the people cannot rule, the strongest party, or combination of parties, should' [565.12]. He opined that coalition governments are less accountable than majority governments.

> The more parties there are, the more difficult it is to form a government. We know this from experience but it also stands to reason. When there are only two parties, a government can easily be formed. But proportional representation makes it possible even for small parties to gain great (often decisive) influence over the formation of a government, and thus even over the government's decisions. (*All Life* 1999: 95; *Alles Leben* 1996: 211)

If one is committed to proportionality and regards it as the 'essence of democracy', which Popper did not, a democrat had to live with the consequences. Popper rejected the view that democracies are based on 'people's governments'. They are run by party governments or rather party leaders' governments.

> For the larger the party, the less united and the less democratic it is, and the less influence those who voted for it have upon the party's leadership and programme. It is wrong to think that a parliament elected by proportional representation is a better reflection of the people and its wishes. It does not represent the people and its views but simply the influence that the various parties (and party propaganda) had upon the electorate on polling day. And it makes it more difficult for polling day to be what it could and should be: a day when the people judge the activity of the government. (*All Life* 1999: 95; *Alles Leben* 1996: 210)

Had Popper lived longer he would have experienced, perhaps to his dismay, that the British system cannot avoid coalitions.[13] A multiple party system has disadvantages, but it is also a reflection of the political spectrum in a country. A smaller party, like the Liberals, may moderate the policies of the dominant partner in a coalition. In the current climate of populism and resurgent nationalism it is noteworthy that Popper argued that major constitutional changes needed a qualified, not a simple majority.[14] Again, he was very much in agreement with republicans on this important constitutional issue. The *vox populi* is a myth, not only because people are not the rulers. It is also a fact that majorities are not always right. Worse, it is a fallacy. The voters' job is to judge the government and leave the governance of a country to qualified politicians. Popper argued that proportional representation had even worse consequences when it came to the removal of incompetent governments.

> First, people know that there are many parties and therefore hardly expect one of them to gain an absolute majority....the people's verdict has not actually been expressed against any of the parties. None of them has been thrown out, none has had judgement passed upon it.

---

[13] David Cameron formed a coalition with the Liberal Democrats, and Theresa May's government needed the support of the Northern-Irish DUP (2016–19).

[14] After the British referendum on EU membership (2016) the Leave campaign won with a majority of 51.4% of the vote.

Second, election day is not expected to be a day when the people judge the government. Sometimes it may have been a minority government, forced to make concessions and unable to do what it considered right; or else it may have been a coalition government for which none of the ruling parties was fully responsible.

Third, if a majority of voters want to throw out a majority government, it may not be possible for them to do it. For even if a party that has so far had an absolute majority (and could therefore be held responsible) loses its majority, it will still most probably be the largest single party under the proportional system, and therefore be able to form a coalition government with one of the smaller parties. In that case, the dismissed leader of the main party will continue to govern against the decision of the majority, by relying upon a small party that may be miles from representing 'the will of the people'. (*All Life* 1999: 96; *Alles Leben* 1996: 211–2)

He opposed democracy in the sense of 'rule by the people', which, where it exists, is 'arbitrary and unaccountable dictatorship'.

A government can and should be accountable to the people. Rule by the people cannot be; it is unaccountable.

He was therefore in favour of '*democratically elected, constitutional government'*. (*All Life* 1999: Preface; *Alles Leben* 1996: Vorwort; italics in original) It must be accountable, but it was quite different from rule by the people.

Popper was given the platform to publicize these ideas. It was an expression of the esteem, in which he was held. After such a productive life, it is no wonder that more honours and awards lay in store.

10.5 More Honours

The German government honoured him with the *Orden pour le mérite* for science and the arts (03/06/80). He was a member from 1980 to 1994. The physicist Friedrich von Weizsäcker, who had defended Heisenberg's uncertainty relations against Popper's objections in 1934, gave the Laudatio. His brother Richard von Weizsäcker, who became President of the Federal Republic in 1984, also corresponded with Popper. Popper sent him a copy of *In Search of a Better World,* and von Weizsäcker congratulated him on occasion of his 85th birthday, praising his engagement for critical-rational dialogue [360.22; 567.11] In addition to his Fellowships at the Royal Society and the British Academy he had become an honorary member of countless foreign Academies and had received many medals and prizes. On the occasion of his ninetieth birthday (28/07/92), the prestigious journal *Foundations of Physics*, established by Wolfgang Yourgrau, published four special issues of invited papers (Volume 21/12, December 1991; Volume 22/1–3, January–March 1992). Some honours were more modest. Bath College of Further Education decided in 1992 to confer its first ever honorary degree on Popper, as Magee informed him (15/11/92) [550.02].

Karl Popper

Popper's official portrait for the *Orden pour le mérite*. Copyright: Archiv Orden pour le mérite. Reproduced by permission of the Archiv

From 1985 to 1986, he was Visiting Professor at the University of Vienna. In 1986 he became Head at the Ludwig Boltzmann Institut für Wissenschaftstheorie, Vienna, and the first Medawar Lecturer at the Royal Society. From that year he was also Senior Research Fellow at the Hoover Institution on War, Revolution and Peace at Stanford. On October 28, 1991, he was awarded an honorary doctorate at Complutense University in Madrid. As part of the celebrations the Rector of Complutense, Juan Sierra, asked Popper to send him a tape and the score of the aforementioned fugue for organ. It was the only composition that had survived. According to Popper's recollection it was performed, for the first time, in 1992, at the *Escorial* Palace, near Madrid. (*All Life* 1999: 152) In fact, the fugue was written in 1921 and performed, at first privately in June 1991, at St George's Chapel, Windsor, and then publicly at Paisley Abbey, Scotland, in August 1992. (The archives contain no further information about the Madrid performance. The orchestral version seems to have had its world premiere in April 1993 [550.02].) The year 1991 was interesting for another reason. His former student Hubert Kiesewetter, who by then had enjoyed a very successful academic career (with fellowships in Britain, France, the US and worldwide invitations) proposed Popper for an honorary doctorate at his home university (*Katholische Universität*, Eichstätt in southern Germany). Kiesewetter had come to the *LSE* (1967–68) after having studied philosophy in Frankfurt. He had been little impressed by Adorno, Horkheimer and Habermas [315.13] but bowled over by Popper's lectures. He started a long-lasting correspondence with Popper, which eventually grew into a personal friendship. He often expressed his admiration for Popper's humanity and philosophy [546.02; 546.03]. Popper was pleased to accept, and the ceremony took place on May 27, 1991 [546.02]. (By then the two men were close enough to visit each other in their respective homes in the company of their spouses.) In his lecture Popper condemned 'the Cynical Interpretation of History'. Several German newspapers expressed an interest in publishing it. Kiesewetter, who negotiated on Popper's behalf, eventually came to an agreement with the *Frankfurter*

Allgemeine Zeitung (FAZ) that they would publish the article for a fee of 5,500 DM. To everyone's surprise the FAZ finally decided not to publish the article because, as they wrote to Popper, it was not ground-breaking.[15] (At that time Melitta Mew sometimes answered letters; she wrote both in German, her native tongue, and English, whilst Kiesewetter almost always wrote in German.)

Many conferences and colloquia were held in his honour: a three-day conference at the *LSE* and a symposium in Beijing in 1980; a year later conferences in Cerisy-la-Salle and Strasbourg (France); throughout the 1980s meetings in Europe and China; in 1994 a conference session on his theory of the social sciences in Mexico City; in 1995 a symposium was held in Prague to celebrate the fiftieth anniversary of the publication of *The Open Society*; and in 2002 centenary conferences were held in Vienna and Christchuch. (Miller 1997: 405–6) In 1992 he also received the *Kyoto Prize* for his contributions to twentieth-century philosophy. In his acceptance speech (November 11) he revisited his intellectual development and told the audience that he had become a philosopher 'by accident'. He was driven to it, perhaps like many other professional philosophers, by challenging problems, which emanated from other fields, in his case physics, biology and the social sciences.

After all the academic honours he had received, one should not forget some of the civic honours which were bestowed on him. In 1982 Queen Elizabeth II invested him with the insignia of a Companion of Honour. In 1992 the *Goethe Institute* honoured him with the *Goethe Medal* for his services to German language and international cultural relations. (From my readings, I can testify that his mastery of his mother tongue remained impressive into his old age.) In the same year (17/09) President von Weizsäcker invited Popper to lunch at the Villa Hammerschmidt in Bonn on occasion of his 90th birthday. Melitta Mew (13/05/92) sent a list of guests of honour Popper wished to invite: Dr. Günter Wächtershäuser and his wife Dorothy, Georg Sibeck and Frau Carmen [of J.C.B Mohr (Paul Siebeck) Verlag], Professor Dr. Hubert Kiesewetter and his wife Renate, as well as Raymond and Melitta Mew and their son Bernard [567.11]. In 1994 the *Central European University*, in Budapest, made Popper the first laureate of the *Open Society Prize*.

The Poppers had regained their Austrian citizenship in 1976. His home city, Vienna, held a symposium on the occasion of Popper's 80th birthday. In 1992, Vienna made Sir Karl an honorary citizen. A year later, December 17, 1993, he received the *Otto Hahn Peace Prize* in Berlin. (As a twelve-year-old, at the beginning of

---

[15] I have translated loosely. When Popper asked for an explanation, Mr. Patrick Bahners of the FAZ answered (July 9, 1991): 'Unsere Erwartung, einen Jahrhundertext zu lesen, wurde (….) falsifiziert' [546.03]. The lecture was published in *Alles Leben* 1996: Chap. 13; *All Life* 1999: Chap. 10.

World War I, Karl had written what in his *Autobiography* he dismissed as a silly peace poem which celebrated the restoration of peace. The 'poem was entitled "*Das Friedensfest*" or in English Celebrating Peace' as he informed readers of an article in *The Times*, April 11, 1987 [565.12].) The Prize is named after Otto Hahn, the German chemist who, with the help of others (Fritz Strassmann, Lise Meitner and Otto Frisch), is credited with having discovered nuclear fission. He was awarded the Nobel Prize for chemistry in 1944. He was also an honorary citizen of Berlin. The *Otto Hahn Peace Medal in Gold* is awarded in memory of Hahn's worldwide involvement in the politics of peace and humanitarian causes, in particular since the dropping of the atomic bombs on Hiroshima and Nagasaki in August 1945.

In his acceptance speech Popper underlined the necessity of peace and emphasized the importance of international co-operation. (*Alles Leben* 1996: Chap. 16; *All Life* 1999: Chap. 13) In accordance with his Kantian cosmopolitanism, he expressed his enthusiasm for the *United Nations* and its predecessor, the *League of Nations*. But Popper used the occasion to tell his audience that the seeds of his cosmopolitanism and his commitment to peace lay earlier than his reading of Kant's *Perpetual Peace* (1795). His elder sister Dora had given the seven-year-old a book by the Norwegian explorer Fridtjof Nansen. After World War I, Nansen was instrumental in the founding of the *League of Nations* and was awarded the Nobel Prize for Peace (1922). His book *Farthest North* (1897) told the story of the *Fram* expedition (from 1893 to 1896), during which the team, led by Nansen, reached a record northern latitude of 86° 14′. The book had a formative influence on the young Popper.

> No other book had such an influence on my childhood. It aroused my interest in discoveries – not only voyages of discovery, but also theoretical discoveries. It was Nansen's book that revealed to me as a child the importance of bold theories, of daring, even too daring, hypotheses. For the planning of Nansen's expedition was based upon theoretical considerations and bold ideas. (*All Life* 1999: 141)[16]

Vienna honoured Popper in other ways. A street is named after him, near the main railway station. (There is also a *Popperstraße* in Berlin, and several streets in the United States bear his name.) Vienna also hosts the *Europäische Volksschule Sir Karl Popper* and the *Sir Karl Popper School* at the *Wiedner Gymnasium*. The *Gymnasium* or grammar school is a public school, located in the 4th district of Vienna. The *Sir-Karl-Popper-Schule* forms an integral part of the *Wiedner Gymnasium*. Its two classes per year are reserved for highly gifted pupils comprising grades 9 to 12. Pupils are admitted if they pass several *IQ* tests.

Finally, a plaque on the side of the entrance to the Guildhall of Viennese Cabinet Makers reminds visitors of Popper's apprenticeship (1922–24) with Adalbert Pösch. (Fig. 10.1).

---

[16] More details of these influences on Popper's ethics can be found in Kiesewetter (1995: 275–78); Shearmur (1996: Chap. IV); Artegas (1999); Hacohen (2000: 511–20).

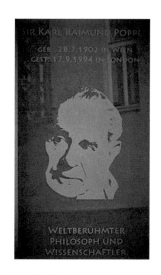

The Guildhall of Viennese Cabinet Makers with a
commemorative plaque for Karl Popper. *Source*:
Wikimedia Commons

Karl-Popper Straße, Vienna
Source: Wikimedia Commons

Memorial Plaque at the house in Anton Langer
Gasse, Vienna, where Karl and Hennie Popper
lived from 1930-35
Source: Wien Geschichte Wiki

Karl-Popper Schule at the *Wiedner
Gymnasium*. *Source*: Wikimedia Commons

Sir Karl-Popper Volksschule. *Source*:
Wikimedia Commons

Fig. 10.1 Vienna honours Sir Karl Popper

Chapter 11
Conclusion

> Thus we are ultimately interested in theories and in their truth, rather than in concepts and their meaning. (K. Popper, *Quantum Theory and the Schism in Physics* 1982/1992: 44)

When people think of Popper what comes to mind is his Critical Rationalism, which comprises his fallibilist methodology and his fight for an open society. But there are two other important aspects, which emerge from his work: his emphasis on situational analysis—the logic of the problem situation—and his insistence on the applicability of his philosophy to areas outside of philosophy. In this intellectual biography I have endeavoured to apply Popper's method of situational analysis to his own work. This method aims at identifying a problem situation. It consists of a *problem*, as it appears to a community of researchers, a toolbox of available *techniques*, with which the problem may be tackled and a number of competing but tentative *solutions*. A problem situation emerges at a particular moment in time against the background of earlier problems and attempted solutions. The history of science bristles with examples: from the Greek investigation of matter to the diversity of species and the physics of the atom. The community agrees that a problem exists, for otherwise it would be ignored as a non-problem. Its members may try to solve the problem with traditional techniques, inherited from the past. The problem may also resist attempts at a solution. Innovative individuals may then invent new methods to get to grips with the problem. Newton and Leibniz independently invented the mathematical technique of calculus to work out such physical properties as volumes and instantaneous velocities. Einstein adopted tensor analysis to deal with the problem of curved space-time.

What was Popper's problem situation? It arose against the background of his critical engagement with Kant's philosophy and logical positivism, the official position of the *Vienna Circle*. (The *Vienna Circle* itself saw metaphysical speculation as a problem, which had to be overcome.) The young Popper was also impressed by Einstein's overthrow of the long-established Newtonian theory of gravitation. This mixed background instilled in him the conviction that the old quest for certainty of knowledge was an illusion and that science did not start with sense data.

© Springer Nature Switzerland AG 2022
F. Weinert, *Karl Popper*, Springer Biographies,
https://doi.org/10.1007/978-3-031-15424-9_11

Popper's whole career was devoted to showing that certain knowledge was beyond human grasp. Not because human minds were too weak to gain knowledge about the external world. No, what the human mind had achieved since Copernicus, or even since the Greeks, was admirable. Prior to Einstein's revolution a belief in the Cartesian certainty of knowledge was considered reasonable. Inductivism was the official methodology of science. But Einstein's achievement—especially the General theory of relativity—demonstrated that all our knowledge claims can be doubted. From this insight Popper drew the conclusion that all knowledge is conjectural. Popper replaced the Cartesian quest for 'certain' knowledge with the model of 'conjectural' knowledge. Still, just like testability, conjectures come in degrees. Popper can be credited with having achieved a paradigm shift. He called it a 'gestalt switch' (Schilpp II: 1044): from certainty of knowledge and trust in authority to conjectures, criticism and testability. He introduced a new 'paradigm': abandon the quest for certainty and replace it by the quest for well-tested hypotheses. His endorsement of *sapere aude*, Kant's motto of the Enlightenment, implied keeping a critical distance from the sanctity of authority, knowledge or traditions. Popper's Critical Rationalism, his trial-and-error method, is a continuation of the Enlightenment project.

The type of scientific theory that Popper had in mind was universal in scope but finite in its experimental or observational data. This meant, for Popper, that scientific theories cannot be confirmed or justified. It also meant that the truth of a scientific theory could not be established on the basis of inductive procedures. Certain truth eludes us. At the time of Einstein's triumphs, Freudianism and Marxism claimed scientific respectability. Freud explicitly stated that psychoanalysis was on a par with physics. So Popper sought a demarcation criterion to distinguish genuine scientific from pseudo-scientific claims. It could not lie in the verification of theories, for this procedure was based on inductive methods. Popper responded to this problem with the introduction of a new technique: falsification, and a new criterion: falsifiability. There existed an asymmetry between verifiability and falsifiability. Popper's solution of falsifiability, I have argued, is itself a tentative solution. Popper never drew a distinction between falsifiability, refutability and testability. Yet there are good reasons to think that falsifiability and refutability are limit cases of a broader notion of testability. Ironically, Popper's tetradic scheme is an illustration of testability, which is an inference to the most plausible explanation. This characterization fits the conjectural nature of knowledge.

His method of falsifiability, partly anticipated by Peirce (*Autobiography* 1974: 1065), later turned into a general attitude, which he dubbed 'Critical Rationalism'. It involves a defence of objectivism (as against subjectivism), the reliance on evidence and above all a critical attitude towards all kinds of authority. Popper changed the image of science and politics. Replace the blind trust in authority and traditions by critical evaluation. Do not submit blindly to any kind of authority—be it scientific, political or social—without careful consideration. This Critical Rationalism is the right kind of attitude in an open society. Kant believed that he lived in an age of Enlightenment but not in an enlightened age. Enlightenment is, in the words of Habermas, an unfinished project. Popper would have agreed. He regarded rationalism and the open society as ideals, which are constantly under review. Popper, as I have emphasized, repeatedly expressed his belief in the power of ideas. Even 'lunatic ideas' had a place and a function in the 'critical method of science'. (*Realism/Aim* 1983/1985: 129) He considered himself as the heir of Kant and the Enlightenment. He was not as great a systematizer as Kant, who wrote major works on epistemology, ethics and aesthetics. But Critical Rationalism is the *leitmotif* of his work. It is the guarantor of the rational growth of knowledge. It is the lifeline of a democracy but the bane of a totalitarian system.

In addition to testability and the open society, Popper's situational logic is the third pillar of his system. It has great value in all areas of inquiry. It requires us to set problems into the context in which they arise. Otherwise it is impossible to understand the solutions. Arguably the most famous philosophical slogan of all time is Descartes's *cogito ergo sum*: I think therefore I am. This Cartesian thesis is impossible to appreciate without realizing that it was an attempted solution to a problem, which exercised the best minds of the seventeenth century. The Cartesian problem was how to acquire reliable knowledge. The prevailing mood at the end of the sixteenth century was scepticism, well captured in Montaigne's *Essais* (1595). It is the view that the human mind is too weak to acquire reliable knowledge of the external world. Understanding remained on the level of mere opinions. Against this background Descartes aimed to show that his famous method of doubt could guarantee *certain* knowledge of the external world. Whilst Descartes regarded his solution as definitive, Popper rejected it as tentative. Popper's achievement is to have replaced the Cartesian paradigm by the model of conjectural knowledge.

A largely unnoticed feature of Popper's philosophy is his increasing emphasis on a structural view of reality. Underlying the appearances, the observable phenomena, lies a 'hidden' reality. It consists of structures. Change is fundamental. Hence the world is a dynamic place. Popper rejected the popular view of a Laplacean block universe. If change is not completely predictable, as Popper argued, an indeterministic worldview is an almost inescapable consequence. This is striking in human affairs: there are patterns of behaviour and trends but no universal laws, as in the physical universe.

It is only right to ask whether Popper's ideas have exerted an influence on his and succeeding generations. His methodology found admirers in a number of prominent scientists (Bohm, Bondi, Eccles, Medawar, Nurse). Politicians such as Helmut Kohl, Helmut Schmidt and Bryan Magee and financiers such as George Soros applauded his fight for an open society. They adopted his Critical Rationalism. It may have lost its influence amongst professional philosophers. But David Miller, Jeremey Shearmur, Hans Albert and Hans-Joachim Niemann keep the flame alive.

Critical Rationalism, testability and anti-authoritarianism constituted his solution to the problem situation, which he had inherited from the *Vienna Circle* and the Cartesian tradition. It covered the natural sciences (physics and evolutionary biology), the social sciences and politics. His solution will, I believe, be of abiding interest. Popper saw in it the 'unity' of his work. (*Autobiography* 1974: 1053)

Bibliography

Note: This bibliography only contains titles cited in this book. They are listed in alphabetical order. Square brackets show the abbreviated titles used in the text. Where available I have quoted both the German and the English texts. Comprehensive bibliographies, including secondary literature, are available in Schilpp II (1974); Miller (1997), Corvi (1997), Hacohen (2000). The Karl Popper Sammlung at the University of Klagenfurt contains a Popper Bibliography in German: https://www.aau.at/universitaetsbibliothek-klagenfurt/karl-popper-sammlung/ and English: https://www.aau.at/en/university-library-klagenfurt/karl-popper-collection/. The Hoover Institute has an online Register of Popper's Papers, Manuscripts and Correspondence, from 1928 to 1995: Karl R. Popper papers (cdlib.org) http://pdf.oac.cdlib.org/pdf/hoover/reg_189.pdf. The German edition of Popper's *Autobiography*, entitled *Ausgangspunkte* (^{5}2018), contains an updated bibliography from 1925 to 2003.

I. *Works by Karl R. Popper* (titles in alphabetical order)

Alles Leben ist Problemlösen [Alles Leben]. München: Piper (1996). English translation: *All Life is Problem Solving* [*All Life*]. London: Routledge (1999)[1]
'A New Interpretation of Darwinism', in Niemann (2014): 115–28
A World of Propensities [*Propensities*]. Bristol: Thoemmes (1990)
Auf der Suche nach einer besseren Welt [*Auf der Suche*]. München: Piper (1984/^{2}1987). English Translation: *In Search of a Better World* [*In Search*]. London/New York: Routledge 1992
Ausgangspunkte. München: Piper (2004) (German translation of *Autobiography* 1974, revised by Karl Popper, includes an updated bibliography)
Autobiography, in: Schilpp, *ed*. (1974), Vol. I: 3–181; published in paperback as *Unended Quest* (see below)
Conjectures and Refutations [*C&R*]. London: Routledge & Kegan Paul (1963)
Die beiden Grundprobleme der Erkenntnistheorie [*Grundprobleme*], hrsg. von Troels Eggers Hansen. Tübingen: J. C. B. Mohr (Paul Siebeck) (1979). English translation: *The Two Fundamental Problems of the Theory of Knowledge,* edited by Troels Eggers Hansen. Abington: Routledge (2009)

---

[1] The English translation omits the following essays from the German original: 'Wissenschaftliche Reduktion und die essentielle Unvollständigkeit der Wissenschaft (Chap. 2); 'Über Geschichtsschreibung und über den Sinn der Geschichte ' (Chap. 8); but it contains 'Masaryk and the open society ' (Chap. 14) and 'How I became a philosopher without trying ' (Chap. 15), which were originally written in English.

© Springer Nature Switzerland AG 2022
F. Weinert, *Karl Popper*, Springer Biographies,
https://doi.org/10.1007/978-3-031-15424-9

'Die Logik der Sozialwissenschaften', in: Th. W. Adorno et al. (1969): 103–23. English translation: 'The Logic of the Social Sciences', in: Th. W. Adorno et al. (1976): 87–104; reprinted in *In Search of a Better World* (1992): Ch. 5

Die Zukunft ist Offen [*Zukunft*]. München: Piper (1985/⁴1990)

'Evolution', Letter to the *New Scientist* (21. 08. 1980): 611

'Freedom and Intellectual Responsibility', in: *Lesson* (1997): 81–91. Translation of 'Freiheit und intellektuelle Verantwortung', in: *Auf der Suche* (²1987): 239–54

Ich weiß, daß ich nichts weiß – und kaum das [*Ich weiß*]. Berlin: Ullstein (1987/²1992)

'Indeterminism in Quantum Physics and in Classical Physics', Part I, *Brit. J. Phil. Sc.* **1**/2 (1950): 117–33; Part II **1**/3 (1950): 173–195

'Interpretation of Nebular Red-Shifts'. *Nature* **145**, 69–70 (1940)

'Irreversibility and Mechanics'. *Nature* **178**: 382 (1956a)

'Irreversible Processes in Physical Theory'. *Nature* **179**: 1297 (1957)

'Julius Kraft 1898–1960'. *Ratio* **4** (1962): 2–12, reprinted in *Popper: After the Open Society* (edited by Jeremy Shearmur/Piers Norris Turner). London: Routledge (2008): 13–25

Knowledge and the Body-Mind Problem [*Knowledge*], *ed.* M. A. Notturno. London: Routledge (1994)

'Letter to the Editor'. *Times Literary Supplement* (26/03/1970): 338–39

Logik der Forschung [*Logik*]. Tübingen: J. C. B. Mohr (Paul Siebeck) (1935/⁵1973). English translation: *The Logic of Scientific Discovery* [*Logic*]. London: Hutchinson (1959/1980)

'Natural Selection and the Emergence of Mind' (Darwin College). *Dialectica* **32**/3–4 (1977): 339–55

Objective Knowledge. Oxford: Clarendon Press (1972)

Offene Gesellschaft – Offenes Universum [*Offene Gesellschaft*]. München: Piper (1982/³1983)

'On theories as nets'. *New Scientist* (29. July 1982): 319–20, reprinted in: *Open Universe* (1982/1988): Ch. III

Postscript to the *Logic of Scientific Discovery*. Volumes I, II, III (edited by W. W. Bartley III, 1983)

- Volume I: *Realism and the Aim of Science* [*Realism/Aim*]. London: Hutchinson (1983/Paperback 1985)
- Volume II: *The Open Universe*: An Argument for Indeterminism. London: Routledge (1982/Paperback 1988)
- Volume III: *Quantum Theory and the Schism in Physics* [*Quantum Theory/Schism*]. London: Unwin Hyman (1982)/Routledge (1992)

Quantum Theory and the Schism in Physics, see Postscript Vol. III

Realism and the Aim of Science, see Postscript Vol. I

'Reflections on the Theory and Practice of the Democratic State', in: *Lesson* (1977): 65–80. English translation of 'Bemerkungen zur Theorie und Praxis des demokratischen Staates', in: *Auf der Suche* (²1987): 215–38

'Replies to My Critics' ['Replies'], in: Schilpp, *ed.* Vol. II (1974): 961–1197

'Structural Information and the Arrow of Time'. *Nature* **214**: 322 (1967a)

'The Arrow of Time'. *Nature* **177**: 538 (1956a)

The Lesson of this Century [*Lesson*]. Abingdon: Routledge (1997)²

The Myth of the Framework, *ed.* M. A. Notturno. London/New York: Routledge (1994)

The Open Society and Its Enemies. Vol. I: *Plato;* Vol. II: *Hegel & Marx*. London: Routledge (1945/⁵1966)

'The Open Society and its Enemies revisited'. *The Economist* (23.04.1988): 25–28

The Open Universe, see Postscript Vol. II

The Poverty of Historicism [*Poverty*]. London: Routledge (1957/²1960)

'The Propensity Interpretation of Probability'. *Brit. J. Ph. Sc.* **10**/37 (1959): 25–42

The World of Parmenides [*Parmenides*]. London: Routledge (1998), *ed.* A.F. Peterson

---

² This volume contains: 'Reflections on the Theory and Practice of the Democratic State' (Chap. 8); and 'Freedom and Intellectual Responsibility' (Chap. 9); both essays form Chaps. 10 and 11 in *Alles Leben* (1996) respectively.

'Time's Arrow and Entropy'. *Nature* **207**: 233–34 (1965)
'Time's Arrow and Feeding on Negentropy'. *Nature* **213**: 320 (1967b)
Unended Quest – An Intellectual Biography. Fontana (1976)
'Zur Kritik der Unbestimmtheitsrelationen'. *Naturwissenschaften* **22**/48 (1934): 807–808
'Zur Philosophie des Heimatgedankens'. *Die Quelle* **77**/10 (1927): 899–908
Popper, K./J. Eccles: *The Self and Its Brain:* An Argument for Interactionism. Heidelberg/Berlin/New York: Springer (1977)

II. *Secondary Literature*

Adorno, Th. (et al.): *Der Positivismusstreit in der deutschen Soziologie.* Neuwied/Berlin: Luchterhand (1969). English translation: *The Positivist Dispute in German Sociology.* London: Heinemann (1976)
Adorno, Th./M. Horkheimer: *Dialektik der Aufklärung.* Frankfurt a.M.: S. Fischer (1947/1969). English translation: *Dialectic of Enlightenment.* Stanford: Stanford University Press (1972/2002)
Adorno, Th. (1976): 'On the Logic of the Social Sciences', in: Adorno et al., *The Positivist Dispute* (1976): 105–122
Agassi, J.: 'Critical Study. Postscript: On the Futility of fighting Philistines'. *Philosophia* **4**/1 (1974): 163–201
Agassi, J.: 'Institutional Individualism'. *The British Journal of Sociology* **26**/2 (June 1975): 144–155
Agassi, J.: *A Philosopher's Apprentice*: In Karl Popper's Workshop. Amsterdam/New York: Editions Rodopi (2008)
Albert, H.: 'The Myth of Total Reason', in: Adorno et al., *The Positivist Dispute* (1976): 163–197
Al-Khalili, J.: *Black Holes, Wormholes, and Time Machines.* Boca Raton: CRC Press (^{2}2012)
Angelidis, Th. D.: 'On Some Implications of the Local Theory Th(g) and of Popper's experiment', in: R.L. Amoroso et al. *eds.*, *Gravitation and Cosmology*: From the Hubble Radius to the Planck Scale. Kluwer Academic Publishers (2002): 525–536
Artigas, M.: *The Ethical Nature of Karl Popper's Theory of Knowledge.* Bern: Peter Lang (1999)
Bacon, F.: *Novum Organum* (1620)
Bernays, P.: 'Concerning Rationality', in: Schilpp, *ed.* Vol. I (1974): 597–605
Bartley III, W.W.: 'Theories of Demarcation between Science and Metaphysics', in: Lakatos, I./A. Musgrave *eds.*, *Problems in the Philosophy of Science.* Amsterdam: North-Holland Publishing Company (1968): 40–64, 102–19
Bartley III, W.W.: 'Rehearsing a Revolution: Music and Politics' (extract from *Karl Popper: A Life*). Paper presented at the Pacific Regional Meeting of the Mount Pèlerin Society, Christchurch, New Zealand (27–30 November 1989)
Beck, U.: *Risikogesellschaft.* Frankfurt a.M.: Suhrkamp (1986). English translation: *Risk Society.* London: Sage Publications (1992)
Bernard, C.: *Introduction à l'étude de la médecine expérimentale.* Paris: Garnier Flammarion (1966/1865)
Bohm, D.: *The Special Theory of Relativity.* London: Routledge (1966)
Bondi, H.: *Assumption and Myth in Physical Theory.* Cambridge: Cambridge University Press (1967)
Bondi, H.: 'The philosopher for science'. *Nature* **358**: 363 (1992)
Born, M.: 'The statistical interpretation of quantum mechanics.' Nobel Lecture (December 11, 1954): 256–67
Born, M.: 'Statistical Interpretation of Quantum Mechanics', *Science* **122**/3172 (14. 10. 1955): 675–79
Braunstein, D. (Hg.): *Theodor W. Adorno und Karl R. Popper: Briefwechsel 1961*, in: *WestEnd.* Neue Zeitschrift für Sozialforschung **16**/2 (2019): 205–219
Bronowski, J.: 'Humanism and the Growth of Knowledge', in: Schilpp, *ed.* Vol. I (1974): 606–31

Brudny, M. I.: *Karl Popper*: Un philosophe heureux. Paris: Grasset (2002)

Campbell, D. T.: 'Evolutionary Epistemology', in: Schilpp, *ed*. Vol. I (1974): 413–63

Carr, E. C.: *What is History?* Penguin Books (1961)

Carroll, S.M.: 'What Scientific Idea is Ready for Retirement?' Published on Edge.org (https://www.edge.org) (2014)

Carroll, S.M.: 'Beyond Falsifiability: Normal Science in a Multiverse', in: *Why Trust a Theory?*, *eds*. R. Dawid, R. Dardashti, K. Thebault. Cambridge: Cambridge University Press (2019): Ch. 16

Corvi, R.: *An Introduction to the Thought of Karl Popper.* Routledge (1997/Taylor and Francis e-library 2005)

Cox, B./J. Forshaw: *The Quantum Universe*: Everything That Can Happen Does Happen. Penguin Books 2012

Coyne, J. A.: *Why Evolution is True.* Oxford: Oxford University Press (2009/Paperback 2010)

D'Abramo, G.: 'The peculiar status of the second law of thermodynamics and the quest for its violation.' *Studies in History and Philosophy of Modern Physics* **43** (2012): 226–235

Darwin, Ch.: *The Origin of Species* (darwin-online.org.uk, 1st edition 1859)

Darwin, Ch.: *The Descent of Man* (darwin-online.org.uk, 1st edition 1871)

De Santo, F.: 'Karl Popper's forgotten role in the quantum debate at the edge between philosophy and physics in 1950s and 1960s.' *Studies in History and Philosophy of Modern Physics* **67** (2019): 78–88

Deutsch, D.: *The Fabric of Reality.* Penguin Books (1997)

Donohue, J. J.; Levitt, S. D.: 'The Impact of Legalized Abortion on Crime'. *The Quarterly Journal of Economics* **116**/2 (2001): 379–420.

Eccles, J.C.: 'The World of Objective Knowledge', in: Schilpp, *ed*. Vol. I (1974): 349–70

Eddington, A.: *The Nature of the Physical World.* Cambridge: Cambridge University Press (1932/1928)

Eddington, A.: *New Pathways in Science.* Cambridge: Cambridge University Press (1935)

Edmonds, D./J. Eidinow: *Wittgenstein's Poker.* New York: Harper Collins (2001)

Einstein, A.: 'Autobiographical Notes', in: P. A. Schilpp, *ed. Albert Einstein – Philosopher-Scientist*, Vol. I: 2–94. La Salle (Ill.): Open Court (1949)

Einstein, A./M. Born: *Briefwechsel 1916–1955.* München: Nymphenburger Verlagshandlung (1969/1991). English translation: The *Born-Einstein Letters.* London: Macmillan Press (1971)

Elster, J.: *Nuts and Bolts for the Social Sciences.* Cambridge: Cambridge University Press (1989)

Epstein, B.: *The Ant Trap.* Oxford: Oxford University Press (2015)

Feigl, H./P. E. Mehl: 'The Determinism-Freedom and Body-Mind Problems', in: Schilpp, *ed*. Vol. I (1974): 520–59

Freeman, E./H. Skolowski: 'The Search for Objectivity in Peirce and Popper', in: Schilpp, *ed*. Vol. I (1974): 464–519

Freud, S.: *An Outline of Psychoanalysis* (1938). In: *The Standard Edition of the Complete Psychological Works of Sigmund Freud* (*ed.* James Strachley). London: The Hogarth Press, Vol. XXIII (1964): 144–207

Friedman, M.: *Essays in Positive Economics.* Chicago: University of Chicago Press (1953)

Futuyma, D.: *Evolutionary Biology.* Sunderland (Mass.): Sinauer (^{3}1998)

Geier M.: *Karl Popper.* Reinbek: Rororo Monographie (1994)

Gillies, D.: 'Popper's Contribution to the Philosophy of Probability', in: O'Hear, *ed*. (1995): 103–120

Gillies, D.: *Philosophical Theories of Probability.* London: Routledge (2000)

Gombrich, E. H.: 'The Logic of Vanity Fair', in: Schilpp, *ed*. Vol. II (1974): 925–57

Gombrich, E. H.: 'Personal Reflections of the Publication of *The Open Society*', in: I. Jarvie, I./S. Pralong, *eds*.: *Popper's* Open Society *After Fifty Years.* Routledge: London (1999): 17–27

Gould, St. J.: *Hen's Teeth and Horse's Toes.* Penguin Books (1984)

Gould, St. J.: *The Panda's Thumb.* Penguin Books (1987)

Gould, St. J.: 'More Things in Heaven and Earth', in: H. Rose/St. Rose, *eds*.: *Alas Poor Darwin.* London: Vintage (2001): 85–105

Gould, St. J.: *The Structure of Evolutionary Theory*. Cambridge (Mass.)/London: The Belknap Press of Harvard University Press (2002)

Grattan-Guinnness, I.: 'Russell and Karl Popper: Their Personal Contacts', in: *Russell*: The Journal of the Bertrand Russell Archives **12** (1992): 3–18

Grattan-Guinnness, I.: 'Karl Popper for and Against Russell', in: *Russell*: The Journal of the Bertrand Russell Archives **18** (1998): 25–42

Greenfield, S.: *A Day in the Life of the Brain*. Penguin Random House UK (2016)

Grünbaum, A.: 'Popper's Views on the Arrow of Time', in: Schilpp, *ed*. Vol. II (1974): 775–97

Habermas, J.: *Theorie des kommunikativen Handelns* I, II. Frankfurt a./M.: Suhrkamp (1981/⁴1987). English translation: *Theory of Communicative Action*. Volume I, Boston: Beacon Press (1984); Volume II, Cambridge: Polity Press (1986)

Habermas, J.: 'The Analytical Theory of Science and Dialectics', in: Adorno et al., *The Positivist Dispute* (1976): 131–162

Hacohen, M. H.: *Karl Popper*. The Formative Years 1902–1945. Cambridge: Cambridge University Press (2000)

Hayek, F. von: 'Scientism and the Study of Society I'. *Economica* **9** (1942): 267–91; reprinted in: Hayek (1964): Part I

Hayek, F. von: 'Scientism and the Study of Society II'. *Economica* **10** (1943): 34–63; reprinted in: Hayek (1964): Part I

Hayek, F. von: 'Scientism and the Study of Society III'. *Economica* **11** (1944): 27–39; reprinted in: Hayek (1964): Part I

Hayek, F. von: *The Counter-Revolution of Science*. New York: The Free Press (1955/1964)

Hayek, F. von: *The Road to Serfdom*. London: Routledge (1944/2014)

Hochkeppel, W.: 'Das Suchen ist die Philosophie'. *Die Zeit* **31** (24. Juli 1987): 31

Horkheimer, M.: *The Eclipse of Reason*. New York: Oxford University Press & New York: The Seabury Press (1947/1974)

Horkheimer, M.: 'Traditionelle und Kritische Theorie' (1937); reprinted in: *Traditionelle und Kritische Theorie*. Frankfurt a./M.: Fischer (1970): 12–64. English translation: *Critical Theory*. New York: The Continuum Publishing Company (1972): 188–252

Huxley, J.: *Evolution: the modern synthesis*. London: Allen & Unwin (1942)

Jablonka, E.: 'The evolutionary implications of epigenetic inheritance'. *Interface Focus* **7** (2017): 1–9

Janik, A./St. Toulmin: *Wittgenstein's Vienna*. A Touchstone Book. New York: Simon & Schuster (1973)

Keuth, H.: *The Philosophy of Karl Popper*. Cambridge: Cambridge University Press (2005)

Kiesewetter, H.: 'Ethical Foundations of Popper's Philosophy', in: O'Hear, *ed*. (1995): 275–288

Kim, Y.-H./Y. Shih: 'Experimental Realisation of Popper's Experiment: Violation of the Uncertainty Principle?' *Foundations of Physics* **29**/12 (1999): 1849–61

Kneale, W. C.: 'The Demarcation of Science', in: Schilpp, *ed*. Vol. I (1974): 205–17

Kraft, V.: 'Popper and the Vienna Circle', in: Schilpp, *ed*. Vol. I (1974): 185–204

Kuhn, T. S.: *The Copernican Revolution*. Cambridge: Harvard University Press (1957)

Kuhn, T. S.: *The Structure of Scientific Revolutions*. Chicago: The University of Chicago Press (²1970/1962)

Kuhn, T. S.: 'Logic of Discovery or Psychology of Research? In: Lakatos/Musgrave (1970): 1–23

Kuhn, T. S.: *The Essential Tension*. Chicago: Chicago University Press (1977)

Lakatos, I./A. Musgrave, *eds.*: *Criticism and the Growth of Knowledge*. Cambridge: Cambridge University Press (1970)

Lakatos, I.: ''Popper on Demarcation and Induction', in: Schilpp, *ed*. Vol. I (1974): 241–273

Laskar, J.: 'A numerical experiment on the chaotic behavior of the Solar System'. *Nature* **338**: 237–238 (1989)

Lecourt, D. (1994): 'Karl Popper: 'ni Marx ni Freud'. *Le Monde* (23 septembre 1994): 13

Levinson, A.: 'Popper, Hume and the Traditional Problem of Induction', in: Schilpp, *ed*. Vol. I (1974): 322–31

Lewontin, R. (1972): 'Testing the Theory of Natural Selection'. *Nature* **236** (24/03/72): 181–2

Liebig, J.: 'Lord Bacon as natural philosopher'. *Macmillan's Magazine* **3** (1863), Pt. I: 237–58, Pt. II: 258–67

Lorenz, K.: *Die Rückseite des Spiegels:* Versuch einer Naturgeschichte menschlichen Erkennens (Piper 1973). English translation: *Behind the Mirror: A Search for a Natural History of Human Knowledge.* New York: Harcourt Brace Jovanovich (1977)

Magee, B.: 'What Use is Popper as a Politician?' In: O'Hear, *ed.* (1995): 259–73

Marcuse, H.: *Reason and Revolution.* New York: Humanities Press (^{2}1954/1941)

Marks, J.: *Relativity.* London: Geoffrey Chapman (1972)

Maxwell, G.: 'Corroboration without Demarcation', in: Schilpp, *ed.* Vol. I (1974): 292–321

Mayr, E.: *What Evolution is.* Basic Books (2001)

Medawar, P.: 'Hypothesis and Imagination', in: Schilpp, *ed.* Vol. I (1974): 274–291

Medawar, P.: *Pluto's Republic.* Oxford: Oxford University Press (1984)

Miller, D.: *Critical Rationalism.* Chicago/La Salle: Open Court (1994)

Miller, D.: 'Propensities and Indeterminism', in: O'Hear, *ed.* (1995): 121–47

Miller, D.: 'Sir Karl Popper'. *Biographical Memoirs of Fellows of the Royal Society* **43** (1997): 367–409; reprinted in D. Miller: *Out of Error.* Ashgate Publishing (2006): Ch. 1

Miller, D.: 'Darwinism is Situational Logic in the State of Ignorance', in: *Karl Popper.* A Centenary Assessment, Volume III (*ed.* I. Jarvie, K. Milford, D. Miller). Aldershot: Ashley 2006: 155–61

Miller. D.: 'Do We Reason When We Think We Reason, or Do We Think? *Learning for Democracy* **1**/3 (2005): 57–71

Miller, D.: 'In Defence of Relative Propensities.' Talk held at the IHPST, Paris (20/06/2008)

Miller, D.: 'Overcoming the Justificationist Addiction'. *Studia Philosophica Wratislaviensia,* Supplementary Volume, English Edition (2012): 93–103

Miller, D.: 'Some hard questions for Critical Rationalism' (2014), available at https://warwick.ac.uk/fac/soc/philosophy/people/miller/

Miller, D.: 'Popper's Contributions to the Theory of Probability and its Interpretation', in: *The Cambridge Companion to Popper*, eds. J. Shearmur/G. Stokes. Cambridge: CUP (2016): Ch. 9

Mulkay, M.: *Science, and Sociology of Knowledge.* London: George Allen & Unwin (1979)

Munz, P.: 'Popper's Contribution to the 20th Century'. *New Zealand Science Review* **48** (1991): 67–71

Munz, P.: 'Great Thoughts and a Liking for Milkshakes'. *The Evening Post* (Wellington, NZ), (September 22, 1994): 5

Munz, P.: *Philosophical Darwinism.* London: Routledge (1993)

Munz, P.: 'Popper's Darwinism', in: *Karl Popper.* A Centenary Assessment, Volume III (*ed.* I. Jarvie, K. Milford, D. Miller). Aldershot: Ashley (2006): 131–41

Musgrave, A. E.: 'The Objectivism of Popper's Epistemology', in: Schilpp, *ed.* Vol. I (1974): 560–96

Nansen, F.: *Farthest North* (1897)

Niemann, H.-J.: *Karl Popper and the Two New Secrets of Life.* Tübingen: Mohr Siebeck (2014)

Nola, R./F. Weinert: 'Darwinian Inferences', in: Brinkworth, M./F. Weinert, *eds.*: *Evolution 2.0.* Heidelberg/Berlin: Springer (2012): 111–28

Noble, D.: 'Secrets of life from beyond the grave'. *Physiology News Magazine* **97** (2014): https://doi.org/10.36866/pn.97.34

Norton, J.: *The Material Theory of Induction.* Calgary: University of Calgary Press (2021)

O'Hear, A.: *Karl Popper.* London: Routledge (1980)

O'Hear, A.: 'Popper and the philosophy of science'. *New Scientist* (22. August 1985): 43–5

O'Hear, A., *ed.*: *Karl Popper*: Philosophy and Problems. Cambridge: Cambridge University Press (1995)

Pais, A.: *Subtle is the Lord.* Oxford: Oxford University Press (1982)

Parvin, P.: *Karl Popper.* London: Bloomsbury Academic (2013)

Penrose, R.: *The Emperor's New Mind.* Oxford: Oxford University Press (1989)

Perutz, M.: 'A New View of Darwinism'. *New Scientist* (October 2, 1986): 36–8

Peterson, I.: *Newton's Clock.* New York: W. H. Freeman and Co. (1993)

Poincaré, H.: *La Science et L'Hypothèse.* Paris: Flammarion (1968/1902)

Poincaré, H.: *Science et Méthode.* Paris: Flammarion (1908)

Polanyi, M.: *Personal Knowledge*. London: Routledge & Kegal Paul (1958)
Putnam, H.: 'The "Corroboration" of Theories', in: Schilpp, *ed*. Vol. I (1974): 221–40
Qureshi, T.: 'Popper's Experiment: A Modern Perspective.' *Quanta* **1** (2013): 19–32
Riley, M.: *Evolution*. Oxford: Blackwell (^{3}2004)
Russell, B.: 'On Denoting.' *Mind* **14** (1905): 479–493
Rutherford, E./F. Soddy (1902): 'The Cause and Nature of Radioactivity I, II'. *Philosophical Magazine* IV, 370–96, 569–85
Schilpp, P. A., *ed*.: *The Philosophy of Karl Popper* Vol. I, II. La Salle (Ill.): Open Court
Schlegel, R.: *Time and the Physical World*. Dover Publications (1968)
Schmidt, H. (1987): 'Der Mann mit dem Goldhelm'. *Die Zeit* **31** (24. Juli 1987): 31
Schorske, C. E.: *Fin-De-Siècle Vienna*. Cambridge: Cambridge University Press (1981)
Schrödinger, E. (1958): 'Mind and Matter', reprinted in *What is Life?* Cambridge: Cambridge University Press (1967): 99–166
Settle, T.: 'Induction and Probability Unfused', in: Schilpp, *ed*. Vol. II (1974): 697–749
Shearmur, J.: *The Political Thought of Karl Popper*. London: Routledge (1996)
Simkin, C.: *Popper's Views on Natural and Social Science*. Brill (1993)
Sober, E.: 'Testability'. *Proceedings and Addresses of the American Philosophical Association* **73** (1999): 47–76
Soros, G.: 'Capital Crimes'. *The Guardian* (January 18, 1997)
Stove, D.: *Popper and After*. Oxford: Pergamon Press (1982)
Suárez, M.: 'Quantum Propensities'. *Studies in History and Philosophy of Modern Physics* **38**/2 (2007) 418–438
Suárez, M.: 'Propensities in Quantum Mechanics', in: Greenberger, D./ K. Hentschel/ F. Weinert, *eds*.: *Compendium of Quantum Physics* (2009): 502–5
Suárez, M.: 'The Chances of Propensities'. *Brit. J. Ph. Sc.* **69** (2018): 1155–77
Suárez, M.: *Philosophy of Probability and Statistical Modelling*. Cambridge: Cambridge University Press 2020.
Susskind, L.: *The Cosmic Landscape*. New York: Back Bay Books (2006)
Suttner von, B.: *Die Waffen Nieder!* (*Lay Down your Arms*) (1889)
Taylor, Ch.: *Hegel*. Cambridge: Cambridge University Press (1975)
Thewissen, J.G.M.: *The Walking Whales*. University of California Press (2014)
Toboso, F.: 'Institutional Individualism and Institutional Change: The Search for a Middle Way of Explanation. *Cambridge Journal of Economics* **25**/1 (2001): 765–83
Vaas, R.: 'Kritische Welterkenntnis. Karl R. Popper und die Kosmologie.' *Aufklärung und Kritik* **1** (2019): 232–53
Watkins, J. W. N.: 'The Unity of Popper's Thought', in: Schilpp, *ed*. Vol. I (1974): Ch. 11
Watkins, J.: Karl Popper: a memoir. *American Scholar* **66** (1997a): 205–219
Watkins, J.: Karl Raimund Popper, 1902–1994. *Proc. Brit. Acad*. **94** (1997b): 645–84
Weber, M.: *Wirtschaft und Gesellschaft*. Tübingen: J. B. Mohr (^{5}1972/1921). English translation: *Economy and Society*. University of California Press (1978)
Weber, M.: '"Objectivity" in Social Science and Social Policy' (1904), reprinted in: *Max Weber on the Methodology of the Social Sciences*, translated and edited by E.A. Shils and H.A. Finch. Illinois: The Free Press of Glencoe (1949): 50–112
Weinberg, St.: *Dreams of a Final Theory*. London: Vintage (1993/1988)
Weinberg, St.: *The Discovery of Subatomic Particles*. Penguin Books (1993/1990)
Weinert, F.: 'Theories, Models and Constraints.' *Studies in History and Philosophy of Science* **30**/2 (1999): 303–333
Weinert, F.: 'The Construction of Atom Models: Eliminative Inductivism and its Relation to Falsificationism'. *Foundations of Science* **5** (2000): 491–531
Weinert, F.: *The Scientist as Philosopher*. Heidelberg/Berlin: Springer (2004)
Weinert, F.: 'The Role of Probability Arguments in the History of Science'. *Studies in History and Philosophy of Science* **41** (2010): 95–104
Weinert, F.: *The March of Time*: Evolving Conceptions of Time in the Light of Scientific Discoveries. New York/Berlin/Heidelberg: Springer (2013)

Weinert, F.: *The Demons of Science:* What they can and cannot tell us about our World. New York/Berlin/Heidelberg: Springer (2016)

Weinert, F.: 'On the Explanation of Social and Societal Facts', in: Wuppuluri, Sh./I. Stewart, eds.: *From Electrons to Elephants and Electrons*: Exploring the Role of Content and Context. New York/Berlin/Heidelberg: Springer (2022): 799–820

White, L. (1992): 'A giant mind winkles wisdom from the small print of history'. *The Sunday Times* (12 July 1992): New Review 8

Whitrow, G. J.: *The Natural Philosophy of Time*. Oxford: Oxford University Press (^{2}1980/1961)

Wild, J.: 'Popper's Interpretation of Plato', in: Schilpp, *ed*. Vol. II (1974): Ch. 29

Williams, R.M.: 'Karl Popper at Canterbury University College – A Personal Recollection'. *New Zealand Science Review* **48** (1991): 75–6

Winzer, K.: 'Physik am Meer'. *Physik Journal* **1** (Januar 2019): 30–34

Worrall, J.: 'Structural Realism: The best of both worlds?' *Dialectica* **43** (1989): 99–124

Worrall, J.: '"Revolution in Permanence": Popper on Theory-Change in Science', in: O'Hear, *ed*. (1995): 75–102

Zouros, G.: *Popper's Experiment and the Interpretation of Quantum Mechanics*. PhD thesis, London School of Economics (2007)

Author Index

A

Adler, Alfred (1870-1937), 13, 23, 27, 36, 52

Adorno, Theodor (1903-1969), 64, 120–123, 125–127, 214

Agassi, Joseph (*1927), 4, 19, 69, 70, 73, 89, 115, 117, 118, 148, 197

Albert, Hans (*1921), 125, 205, 221

Al-Khalili, Jim (*1962), 40

Aristarchus of Samos (c. 310-c. 230 BC), 35

Aristotle (384 BC - 322 BC), 43, 56, 60, 61, 127, 139

Arndt, Arthur (1884-?), 206

Ayer, Alfred J. (1910-1989), 4, 19, 21, 92, 93, 141

B

Bacon, Francis (1561-1626), 37, 51–53, 146

Baldwin, James Mark (1827-1908), 175, 176, 178, 188

Bartley, William III (1934-1990), 4, 7, 26, 43, 88, 115, 117, 149–151, 153, 170, 199, 200

Beck, Ulrich (1944-2015), 33

Berlin, Isaiah (1909-1997), 21, 25, 26, 93, 94, 114, 125, 128, 155, 196, 205, 215, 216

Bernard, Claude (1813-1878), 53

Bohm, David (1917-1992), 39, 196, 221

Bohr, Niels (1885-1962), 54, 90, 102

Boltzmann, Ludwig (1844-1906), 23, 167–169, 214

Bolzano, Bernard (1781-1848), 144

Bondi, Hermann (1919-2005), 39, 194, 221

Born, Max (1882-1970), 102, 104, 106, 108

Bouveresse, Jacques (1940-2021), 196, 204

Boyle, Edward (1923-1981), 43, 57

Brahe, Tycho (1546-1601), 43, 45, 161

Bronowski, Jakob (1908-1974), 19, 20, 31, 48

Bühler, Karl (1879-1963), 16, 17, 54, 185

Bunsen, Robert (1811-1899), 41

Burgess, Tyrrell (1939-2009), 203

C

Cameron, David (*1966), 211, 212

Campbell, Donald T. (1916-1996), 188

Carnap, Rudolf (1891-1970), 24–26, 34, 54, 119, 149, 196, 197

Carroll, Sean (*1966), 38–41, 50, 53

Cassirer, Ernst (1874-1945), 164

Chamberlin, Thomas (1843-1928), 34

Comte, Auguste (1798-1857), 14, 62–64, 66, 77

Copernicus, Nicolaus (1473-1543), 27, 31, 34, 35, 45, 78, 144, 220

Crick, Francis (1916-2004), 194

Crosland, Anthony (1918-1977), 57

D

Darwin, Charles (1809-1882), 2, 3, 8, 27, 34, 50, 51, 64, 107, 115, 130, 131, 137, 142, 144, 148, 173–175, 177, 183, 185, 186, 188, 190

Descartes, René (1596-1650), 153, 156, 182, 187, 221

Dilthey, Wilhelm (1833-1911), 62

Droysen, Johann Gustav (1808-1884), 62

Durkheim, Emile (1858-1917), 69

© Springer Nature Switzerland AG 2022
F. Weinert, *Karl Popper*, Springer Biographies,
https://doi.org/10.1007/978-3-031-15424-9

Subject Index

© Springer Nature Switzerland AG 2022
F. Weinert, *Karl Popper*, Springer Biographies,
https://doi.org/10.1007/978-3-031-15424-9

.

.

Printed in the United States
by Baker & Taylor Publisher Services